purposeful design

Science

Level One

student booklets

purposeful design Science
Level One

Executive Editor
Derek Keenan

Managing Editor
Steven Babbitt

Editing Team
Bonnie Church
Mary Corcoran
JoAnn Keenan
Frieda Nossaman

Design Team
Dan Schultz
Sarah Snider
Shelley Webb

student booklets

purposeful design®
publications
A Division of ACSI
Colorado Springs, Colorado

Purposeful Design Publications is the publishing division of the Association of Christian Schools International (ACSI) and is committed to the ministry of Christian school education, to enable Christian educators and schools worldwide to effectively prepare students for life. As the publisher of textbooks, trade books, and other educational resources within ACSI, Purposeful Design Publications strives to produce biblically sound materials that reflect Christian scholarship and stewardship and that address the identified needs of Christian schools around the world.

Photographs of insects on pages 53, 57, 61, and 63 used by permission of the John May Museum Center, Colorado Springs, Colorado.
Photograph of water strider on page 61 used by permission of LarvalBug: http://home.att.net/~larvalbug/.
Photograph of the fossilized wasp on page 61 by Mary Corcoran and used by permission of the Florissant Fossil Beds National Monument.
Food Guide Pyramid for Young Children on page 151 provided by the U.S. Department of Agriculture.
Photographs of orange trees on page 161 provided by the Citrus Research and Education Center, University of Florida.
Photographs of the lightning, tornado, and hurricane on page 172 provided by the National Oceanic and Atmospheric Administration/Department of Commerce.
Photographs of Doppler radar on page 173 provided by the National Weather Service, Pueblo, Colorado.
Photographs of space-related images on pages 184, 187, 188, and 189 courtesy of NASA.
Photograph of Zebra anglerfish (frogfish) on Transparency T6 Amazing Fish used by permission of CSIRO Marine Research, Australia.

Printed in the United States of America
16 15 14 13 12 11 7

Purposeful design science, level one
Purposeful Design Science series
ISBN 978-1-58331-203-2 Student edition Catalog #7503

Purposeful Design Publications
A Division of ACSI
PO Box 65130 • Colorado Springs, CO 80962-5130
Customer Service: 800-367-0798 • www.acsi.org

purposeful design
Science

Purposeful Design Publications is deeply grateful to the faculty and staff of Briarwood Christian School in Birmingham, Alabama, for the valuable and insightful contributions they have made to the structure and content of the Purposeful Design Science series.

Table of Contents

Discovering by the Scientific Method

Use these skills step by step to find answers to questions.

Problem

Will both kinds of corn kernels pop when heated?

popping corn feed corn

Predict

I think both kinds of corn kernels will pop when heated.

Plan

Experiment

Heat both kinds of corn kernels.

Observe

Your senses help you to observe and find answers.

Your help you hear. Your help you see. Your helps you smell.

Your help you touch. Your helps you taste.

Collect the data

Popping corn does pop when heated.

Feed corn does not pop when heated.

	does pop	does not pop
popping corn	✓	
feed corn		✓

Conclusion

Decide if your prediction is right or wrong.

Both kinds of corn kernels do not pop when heated.

Senses

Use your senses to discover what is around you. Your senses are touching, tasting, hearing, seeing, and smelling.

Your skin helps you feel the smooth blanket.

Your skin helps you feel the rough sandpaper.

Your tongue helps you taste the sweet cookie.

Your tongue helps you taste the salty pretzel.

Your ears help you hear the loud band.

Your ears help you hear the quiet rabbit.

Your eyes help you see the tree is near.

Your eyes help you see the tree is far away.

Your nose helps you smell a fresh flower.

Your nose helps you smell a stinky shoe.

In this book, you will use your five senses to explore the design of God's creation.

All fish have fins.

My Booklet about Fish

Name ______________________________

Connect the dots.

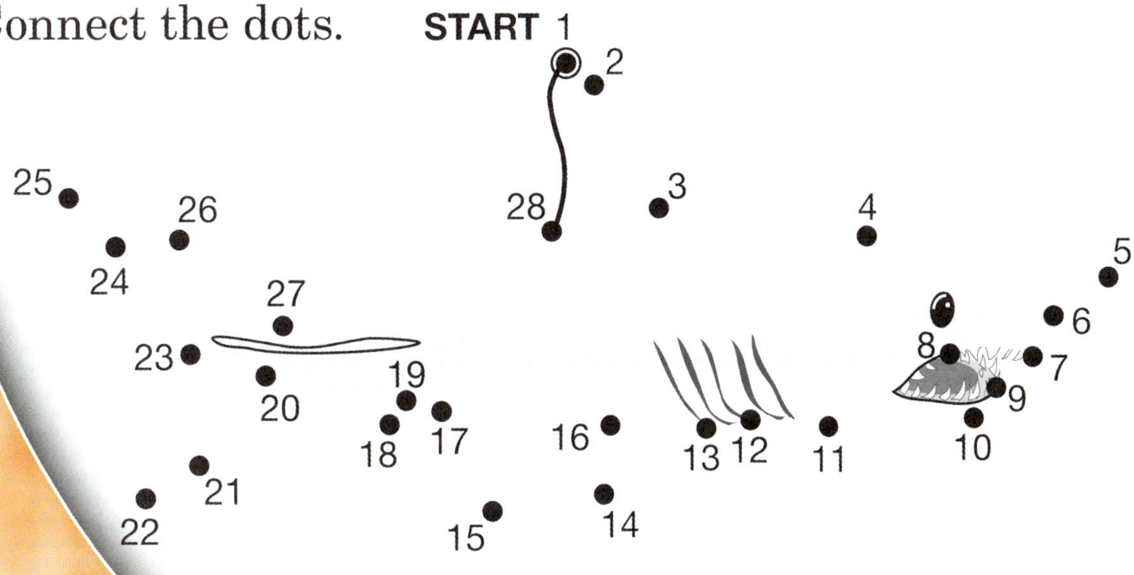

Fish are a special group of animals that live in the water. Fish have fins and most fish have gills and scales.

How do fins, gills, and scales help the fish to live in the water?

Fish use their fins to swim.
Fish use gills for breathing.
Scales protect the body of the fish.

Draw the fins and scales on this fish.

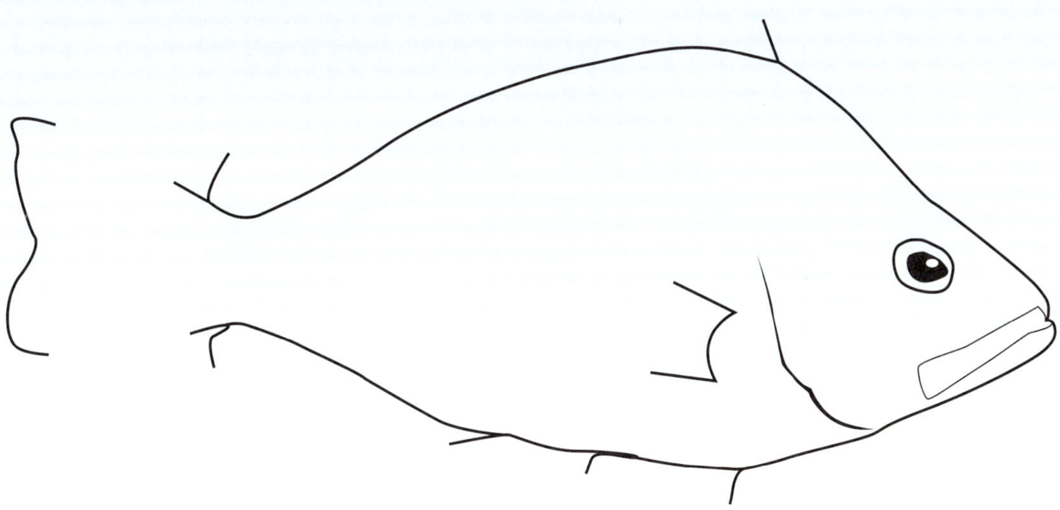

What Are Fish?

A school of fish is a large number of one kind of fish swimming together. A big fish looking for food can be scared off by a large number of fish swimming together.

How many fish do you think are in this school? ______

Lesson Review

1. How are fish alike? Name one way. ______
2. Draw a line to match each body part with how it helps the fish.

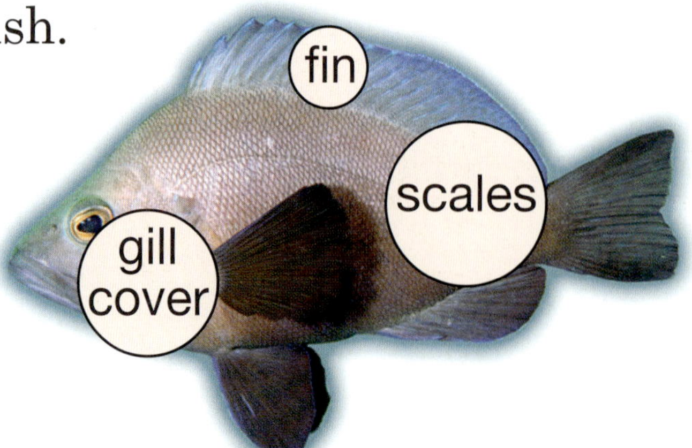

breathe swim protection

How does this fish move in the water?

The dorsal fin helps this fish to stay right side up. The tail fin moves this fish forward. Other fins provide balance and allow the fish to turn in any direction.

How does a fish feel to you when you touch it? Fish may feel slimy. Mucus on the scales makes a fish slippery and helps it swim smoothly through the water. The mucus also keeps germs from getting under the scales.

See how God designed this amazing fish—the sea horse! Its shape is different from many fish. Because of its different shape, its fins have different "jobs."

Can you find a fin?

Parts of a Fish

What body parts help some kinds of fish to sense what is around them?

Fish have eyes but no eyelids.

Their nostrils are used for smelling, not breathing.

Fish use their mouths and outside parts of their body for tasting.

They hear sounds. Their ears are hidden inside their body.

Lesson Review

Word Bank		
dorsal fin	smell	senses

Fill in the blanks.
Use the Word Bank.

1. Nostrils help fish to ______________ things in the water.

2. The ______________ ______________ helps most fish swim and stay right side up.

3. A fish uses its ______________ to get food and stay safe.

How are fish different?

Size Circle the larger fish.

goby

whale shark

Shape Circle the fish that has *lion* as part of its name.

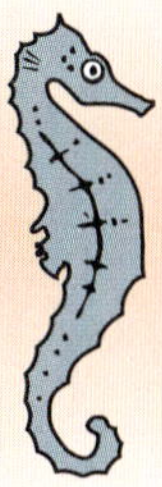
sea horse

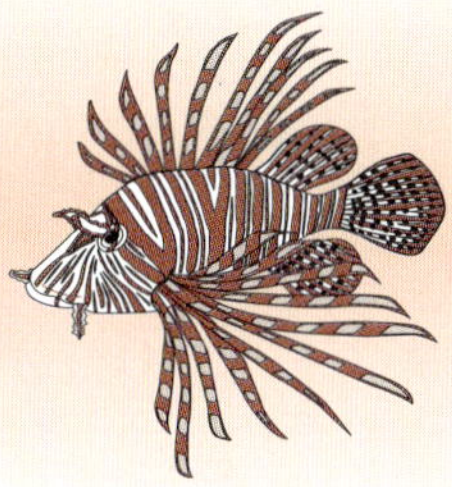
lionfish

angelfish

Color Circle the fish whose name is not a color.

red snapper

walleye

yellow perch

Pattern Circle the fish that has black stripes.

rainbow trout

bluegill

moorish idol

Differences in Fish

A skeleton helps a fish swim. Some fish have a skeleton made mostly of bones.

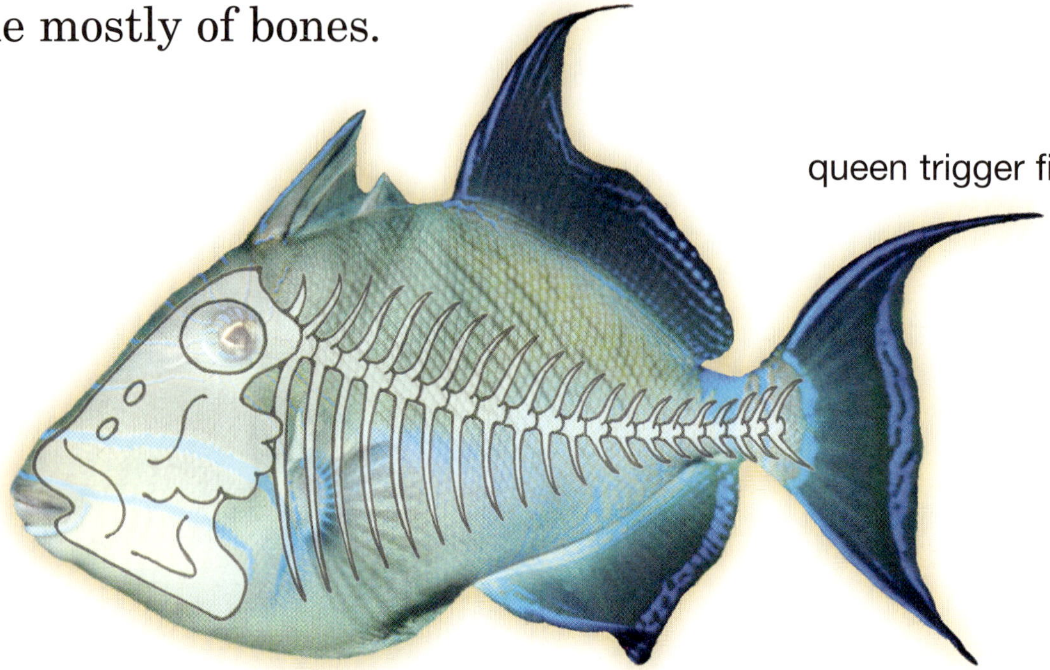

queen trigger fish

Some fish have a skeleton made of tissue called cartilage.

silky shark

stingray

Lesson Review

What are three ways that fish may be different?

1. ______________________________

2. ______________________________

3. ______________________________

Most fish lay eggs. What happens to the eggs?

This clownfish lays its eggs in the coral reef.

Sea horse mothers lay their eggs and put them in the father sea horse's pouch. When the eggs hatch, the baby sea horses come out of the father's pouch.

The mother makes between 200 and 600 eggs at a time. The sea horse is about one centimeter long when hatched.

|—| = 1 centimeter

 Purposeful Design Publications

shark egg case

Some sharks lay their eggs in a tough case where the baby shark grows. The mother leaves the case in the sea. This egg case is known as a mermaid's purse.

How do some fish grow up?

Fish change as they grow up. See how this fish changed from an egg, to a young fish, and to an adult.

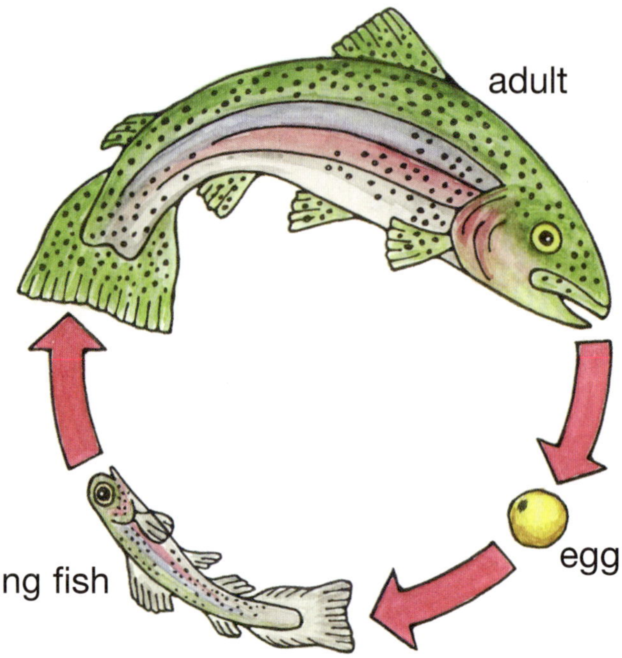

Lesson Review

Fill in the blanks. Use the Word Bank.

Word Bank	
adult	eggs

Some fish hatch from ________________ that grow up to become ________________ fish.

Try this! Observe.

Use a hand lens to observe the table salt.

Draw what you see.

Compare the water in the two containers.

Do the saltwater and freshwater look the same? Do they smell the same? Do they feel the same?

Have you ever been fishing? If so, was it in saltwater or freshwater?

Saltwater habitats are found in oceans and seas. The tarpon is one fish that God designed to live in saltwater.

Habitats and Food

Freshwater habitats are found in lakes, rivers, and ponds. The carp is one fish that God designed to live in freshwater.

What do fish eat?

clownfish

Some fish eat plants.

eel

Some fish eat other animals.

Lesson Review

1. What are two kinds of water God designed fish to live in?
2. What are some things fish eat?

Some fish are designed to blend in with their surroundings. This is a way to hide so they will not be eaten. It is also a way to be unnoticed so they can catch food.

Which fish?
Write the number of each picture on a blank.

_____ My stripes, dots, and color help me to get food.

_____ A red pattern is on my body and in my habitat.

_____ My colors match the sand and plants.

1

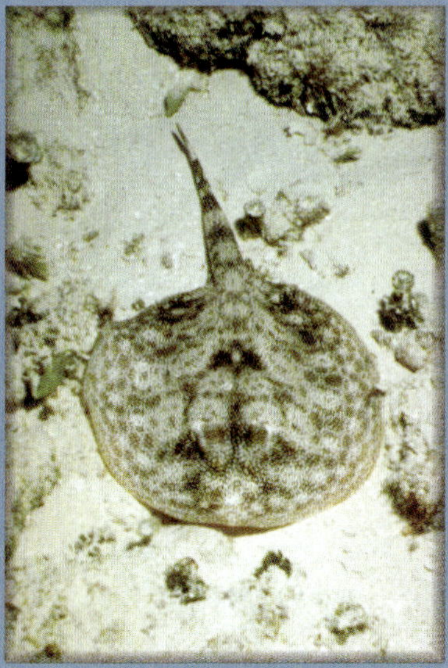

spotted flounder

2

fish in coral reef

3

hawkfish

Make me camouflage!

Why is it better for me to be green than blue?

Camouflage and Defense

These fish are designed to protect themselves from harm. Circle the words that tell what each fish uses as defense.

A butterfly fish has a false eyespot near its tail to trick a predator.

A pufferfish gets spit out of a predator's mouth because it puffs up.

The lionfish uses its poison to sting a predator.

The anemonefish makes mucus to protect itself from the sea anemone's poison.

Lesson Review

How does the design of some fish help them to stay alive?

Write *scales*, *fins*, and *gill cover* to complete each blank.

1. ____________________

2. ____________________

3. ____________________ ____________________

4. Fish use their ____________________ to swim.

5. ____________________ protect the body of a fish.

Write *yes* or *no*. Are all fish alike? ____________________

Draw a fish. Think of size, shape, color, and pattern.

 Purposeful Design Publications

Read the creation web.

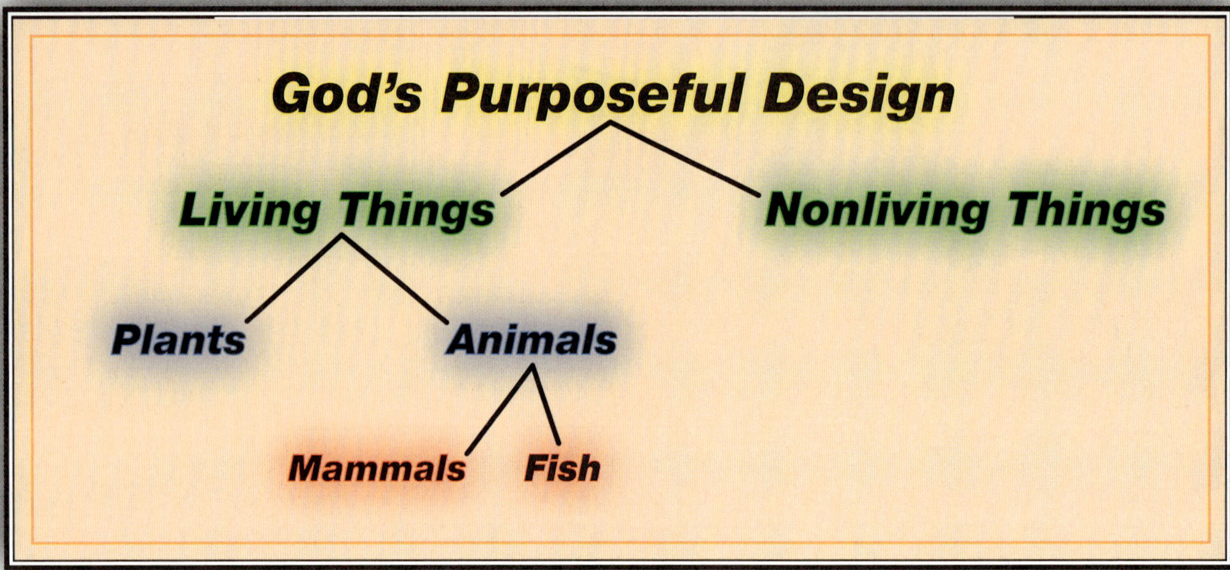

Journal about a Fish

The biggest fish I have ever seen is a

I saw it

Birds have feathers.

My Booklet about Birds

Name ______________________________

Finish coloring the eggs.

"Look at the birds of the air; they do not sow or reap or store away in barns, and yet your heavenly Father feeds them. Are you not much more valuable than they?" Matthew 6:26

Bird-watching can be done in your backyard.

Observe. Draw the place for each bird in the chart.

Bird	Where am I?
 robin	
 blue jay	
 goldfinch	

What Are Birds?

Birds are a special group of animals with feathers and wings. They lay eggs.

Put a check on the line by any of these birds you have seen.

Lesson Review

Circle each word that tells something about birds.

fly	fin	feathers	shoe
bird-watching	animals	pen	wings

Can you identify each part of the blue jay?

Do you know the bird is the only animal with feathers?

Feathers protect their skin and keep them warm or cool.

gray jay

Feathers provide camouflage.

American bittern

Feathers attract other birds.

parrots

Feathers and wings help a bird to fly.

swan

Parts of a Bird

Feathers are different sizes, shapes, and colors.

blue jay flight feather

goose down feather

Birds breathe with lungs. They have a skeleton made of hollow bones.

The bird's light hollow bones help it to fly.

Word Bank

warm
fly
animals
bones

Lesson Review

Use the Word Bank to fill in the blanks.

1. Birds are the only ______________ with feathers.

2. Feathers help a bird to ______________ and to keep ______________.

3. The hollow ______________ of a bird help it to fly.

What are the utensils that help people to eat different foods?

How do birds eat? They have no teeth. God designed different kinds of birds with different beaks.

The beak of a sparrow is designed to crack seeds.

The beak of a spoonbill is designed to sweep from side to side to catch shrimp and snails.

The blue heron uses its beak for stabbing fish.

The hummingbird sticks its beak deep into the flower to get nectar.

Beaks and Food

The hawk's hooked beak tears small mammals, shellfish, and insects for food.

The woodpecker uses its hard beak to drill holes into trees to get insects.

The mallard strains water through its beak for plants and animals.

Fill in the blank.

Is the sparrow designed to eat fish, frogs, nectar, or seeds? ______________

Lesson Review

Write *yes* or *no*.

1. Do beaks on all birds look the same? ______________

2. Do all birds eat the same kind of food? ______________

3. Does the beak of a bird help it to stay alive? ______________

What these people are wearing helps them to do certain things.

God designed different kinds of birds with different feet. Feet help birds to walk, run, climb, swim, perch, grasp, and hop. Their feet are designed for their way of life.

Observe these feet.

Feet

The woodpecker's feet help it to climb.

The puffin's feet help it to swim.

The webbed feet of a goose help it to swim.

The bald eagle's feet help it to grasp.

Many birds use their feet to perch.

parrot

sunbird

Circle the word in each sentence above that tells how each bird uses its feet.

Lesson Review

1. How do webbed feet help the goose and the puffin?
2. Name another bird that has webbed feet.

Even the sparrow has found a home, and the swallow a nest for herself, where she may have her young. Psalm 84:3a

The habitat provides the things a bird needs to live. What birds are found in these habitats? Cut on the dotted lines and place each bird where it was designed to live. Glue.

What is the smallest bird?
What is the largest bird?

This hummingbird is about two and one-half inches long. It weighs less than one ounce. It is the smallest bird.

The ostrich is eight feet tall and weighs 300 pounds. It is the largest bird.

Complete the blanks.

1. The largest bird is the

_______________________________.

2. The smallest bird is the

_______________________________.

Birds Grow

How do birds grow? Birds lay eggs. The eggs hatch. Young birds grow into adults.

Most birds build nests for their eggs and to care for their young.

Observe the place where each bird has its nest.

loon

woodpecker

meadowlark

killdeer

Draw a line from each bird to the place it makes a nest.

1. loon	in a tree
2. woodpecker	near the water
3. meadowlark	near dirt or rocks
4. killdeer	in the grass

Birds Grow

Some birds stay in the same place all year long.

blue jay

cardinal

nuthatch

Some birds fly to a warmer climate for the winter. A move from one place to another is called **migration**. Birds migrate to stay warm and to find food.

arctic tern

sandhill cranes

Lesson Review

1. Why do birds need nests?
2. Why do birds migrate?

Tech Connect

People have learned about flight by studying birds.

1783

The Montgolfier Brothers built a hot air balloon.

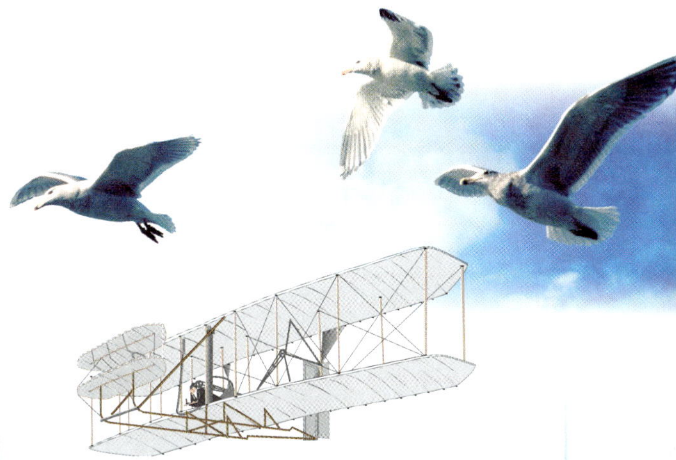

1903

The Wright Brothers flew their aircraft, the *Flyer*.

1927

Charles Lindbergh flew the *Spirit of St. Louis* across the Atlantic Ocean.

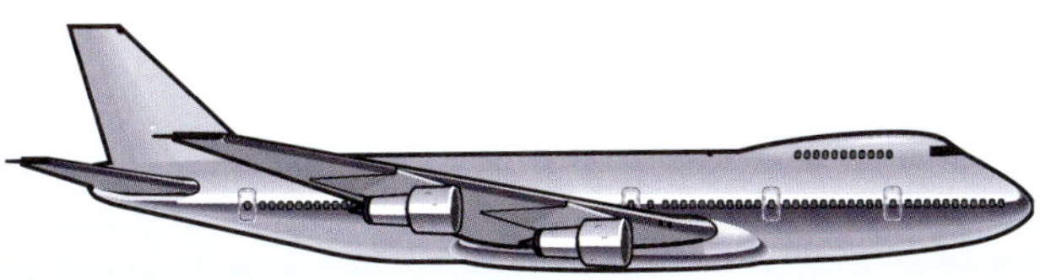

1969

The Jumbo Jet 747 made its first flight.

What is in the sky now?

1981

The first space shuttle mission occurred. Cut on the dotted line and lift the flap.

Tech-Connect

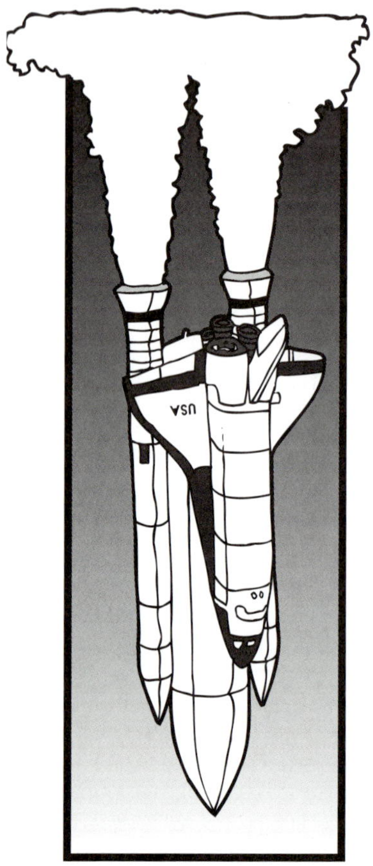

Number the circles 1, 2, and 3 to tell how a bird grows.

grebe

Fill in the circle beside the word that completes the sentence.

1. All ______________ have feathers.

   ○ mammals ○ fish ○ birds

2. The ______________ of a bird's feet helps it to live in its habitat.

   ○ taste ○ shape ○ shoes

3. The ______________ of each bird is special so it can live in its habitat.

   ○ design ○ name ○ hat

4. A bird moves to a warmer climate during its ______________.

   ○ migration ○ eggs ○ feathers

"A mustard seed,...becomes a tree, so that the birds of the air come and perch in its branches." Matthew 13:31b–32b

Chapter 3 Review

Read the creation web.

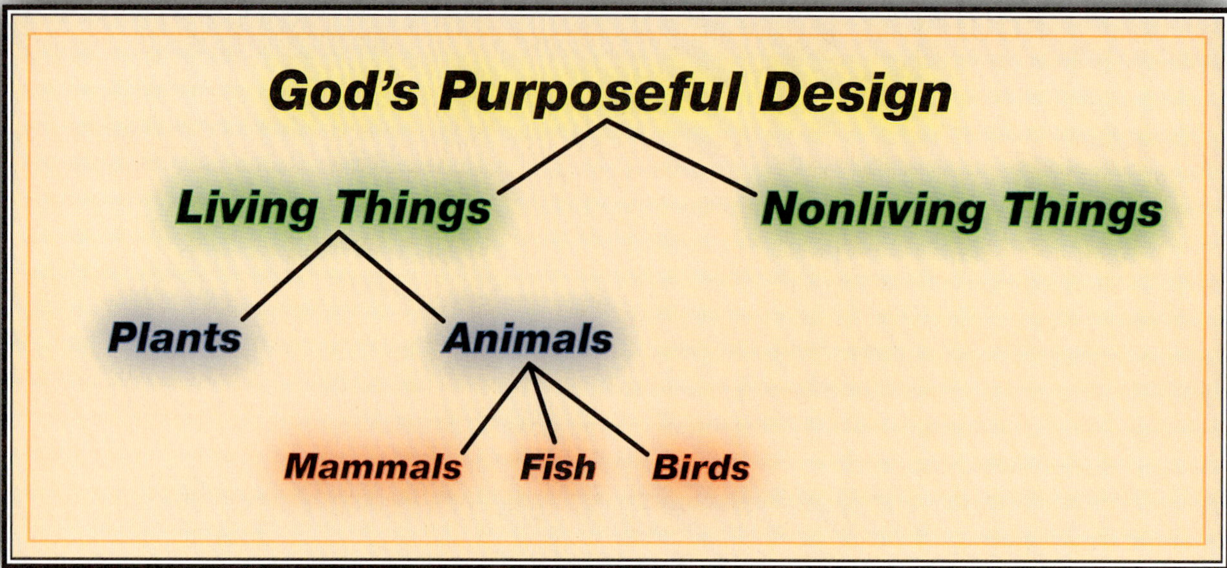

Journal about a Bird

Insects have three main body parts and six legs.

My Booklet about Insects

Name ____________________

Color.

Insects are a special group of animals that have three main body parts and six legs.

Observe the beauty these animals add to the earth.

ladybug beetle

butterfly

dragonfly

moth

What are the three main body parts of an insect? The head, thorax, and abdomen are the three main body parts of an insect.

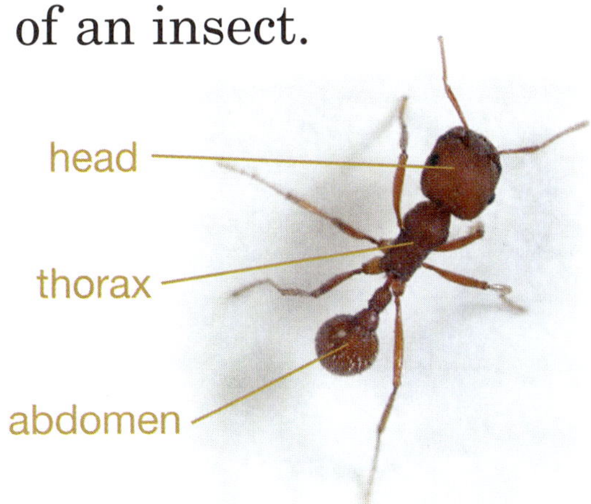

How many legs does this ant have?

____

Ants are creatures of little strength, yet they store up their food in the summer. Proverbs 30:25

Use *head*, *thorax*, and *abdomen* to label the three main body parts of the damselfly.

1. ____________

2. ____________

3. ____________

Is a spider an insect? Observe this spider.

Write *yes* or *no*.

Does the spider have six legs? ____________

Is a spider an insect? ____________

Lesson Review

Word Bank
body
six
three

Use the Word Bank to fill in the blanks.

1. How many main body parts does an insect have? ____________

2. How many legs does an insect have? ____________

3. An insect's ____________ has three main parts.

Where are the insect's eyes, mouth, antennae, and legs?

- Find the eyes, mouth, and antennae on the **head**, the first part.
- Find the six legs connected to the **thorax**, the middle part.
- Find the **abdomen**, the third part.

The insect's outer skeleton is hard and protects the body. It has no backbone. If an insect has wings, they are connected to the thorax.

Make a model of an insect.

Materials
marshmallows
cotton swabs
pencil

Skills
observe
follow a pattern

Follow the steps.

1. Line up three marshmallows.
2. Use a cotton swab to connect three marshmallows.
3. Poke three swabs on both sides of the middle marshmallow. Bend each leg.
4. Add two swabs for the antennae.
5. Draw eyes and a mouth.

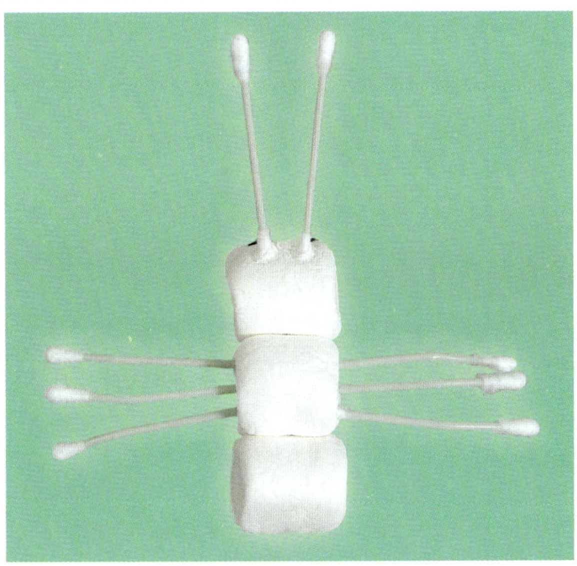

Parts of an Insect

Read each word and identify that part on the insect. Draw a line from the word to the part.

abdomen

head

thorax

antennae

Lesson Review

Circle all the words that tell about a part of an insect.

eyes shirt mouth wings pen

thorax paper football crayon grass

Differences in Insects

The mouth of some beetles helps them to catch prey. It works like pliers.

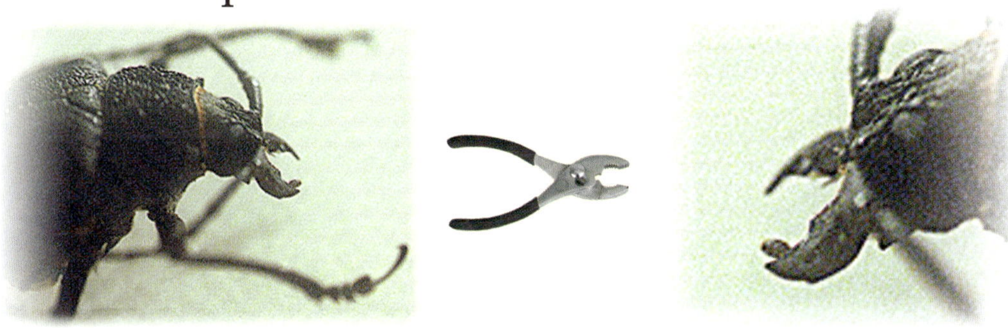

The butterfly's mouth helps it to get nectar from a flower. It works like a straw. It also curls.

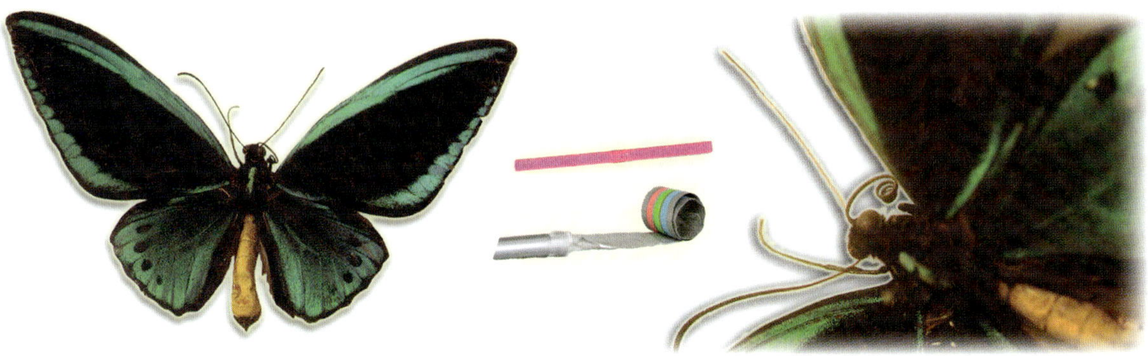

Use a pencil to trace over the close-up photo of each insect's mouth.

Are the mouths of all the insects alike? Circle yes or no.

The mouth of each insect is designed to catch and eat food.

Differences in Insects

The antennae of a butterfly are used to find flower nectar.

The antennae of a moth are used to find other moths.

Are the antennae of all the insects alike? Circle yes or no.

The praying mantis uses its big front legs to catch food.

The grasshopper uses its big hind legs to hop.

Are the legs of all the insects alike? Circle yes or no.

A fly has one pair of wings.

A dragonfly has two pairs of wings.

If an insect has wings, is it always the same number of wings? Circle yes or no.

Lesson Review

Are the mouths, antennae, legs, and wings of all the insects alike? Circle yes or no. Explain.

How do insects grow up?

Insects change as they grow up. See how this butterfly changes.

As insects grow up, they go through metamorphosis, or stages of growth. Their shape is different in each stage.

Number the pictures above 1 through 4 to show the order in which this butterfly grows. Start with the egg.

Graphing

A graph is used to show and compare information that has been collected.

Tim took pictures of insects in different stages of metamorphosis. Observe Tim's photo collection. Color one space on the graph for each photo showing that stage of growth.

Tim's Photo Collection

How many?

Lesson Review

Draw one of the stages of growth of an insect. Name the stage of growth.

Observe each insect.

Use the Word Bank to complete each sentence.

water strider

This insect is so light that it can glide and walk on top of the

____________________.

leafwing butterfly

This insect is designed to look like a

____________________.

Word Bank
leaf
wings
water
wasp

fossilized wasp

This is a fossil of a

____________________.

beetle

A pair of hard wings protects a pair of soft

____________________.

Circle the better answer.

What do the big spots on this insect look like?

eyes or legs

How does the design of this beetle help it?

roll or camouflage

What does this stink bug use as a defense?

smell or sound

What do the hind legs of this insect from South America look like?

leaves or balls

Which insect communicates at night with light?

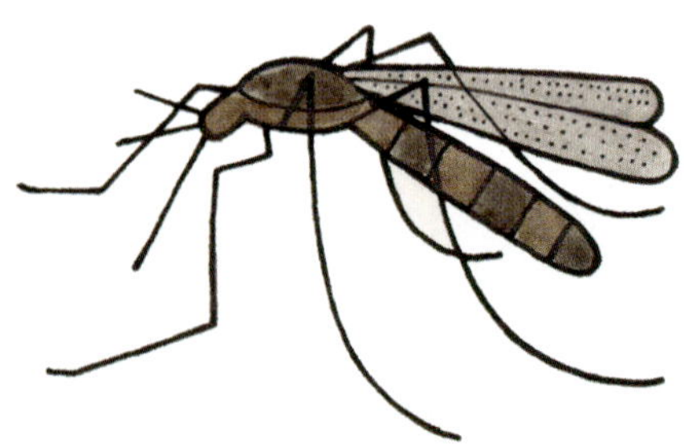

firefly or mosquito

Circle each correct answer.

1. What is the number of main body parts of an insect?

five six three

2. How many legs does an insect have?

six four eight

3. Do the legs of all the insects look alike?

yes no

4. Which stage of growth of a butterfly comes after the chrysalis?

egg adult caterpillar

5. In which habitat would you most likely find a water strider?

Chapter 4 Review

Read the creation web. Write the four groups of animals that the web tells us God purposefully designed.

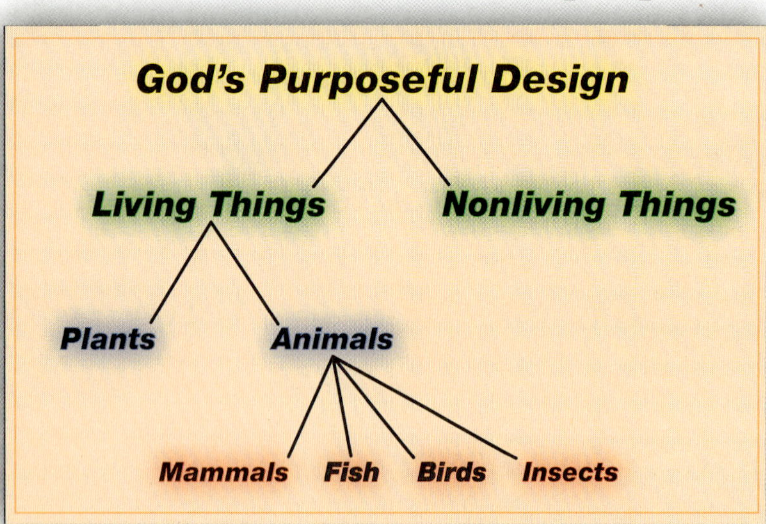

1. ______________________

2. ______________________

3. ______________________

4. ______________________

Journal about an Insect

I learned ____

____

____

____

____.

Wind spins the whirligig.

My Booklet about Movement

Name ______________________________

Color.

When something changes place or position, it moves. This is called movement.

The frog leaps.

It moves from one place to another by leaping.

Write a word from the Word Bank to name a movement in each picture.

Word Bank		
fly	run	spin
kick	slide	jump

What Is Movement?

Movement is all around us. Make a list of any movement you observe in this picture.

Lesson Review

Circle the word that names a movement in each picture.

eat sleep run

jump swim read

Things move in different ways.
Draw how each object moves.

A rocket moves straight up.

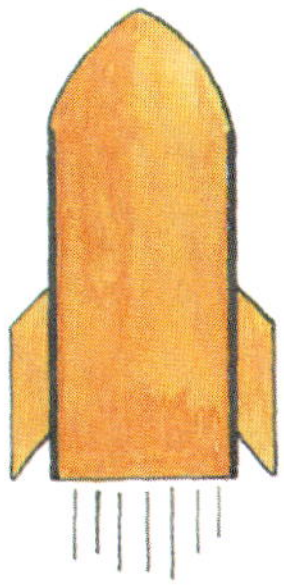

Draw the way the rocket moves.

This train moves round and round.

Draw the way the train moves.

A car on this road moves zizag.

Draw the way the car moves.

A swing moves back and forth.

Draw the way the swing moves.

Some Ways Things Move

Push and pull are forces that can move things.

The boy's push causes the swing to move.

The girl's pull causes the wagon to move.

What is the force that moves the swing?
What is the force that moves the wagon?

Wind causes parts of the plant to move.

A force called gravity causes the apple to fall to the ground.

What causes parts of the plant to move?

What causes the apple to fall to the ground?

He established the force of the wind. Job 28:25a

Lesson Review

Fill in the circle beside the answer.

The swing moves back and forth because

○ of a force
○ it moves by itself
○ of a sound

When something moves back and forth very fast, it vibrates.

What vibrates in each picture? Use the Word Bank.

Word Bank	strings	tuning fork
	triangle	doorstop

The guitar has strings. The strings vibrate when they are played.

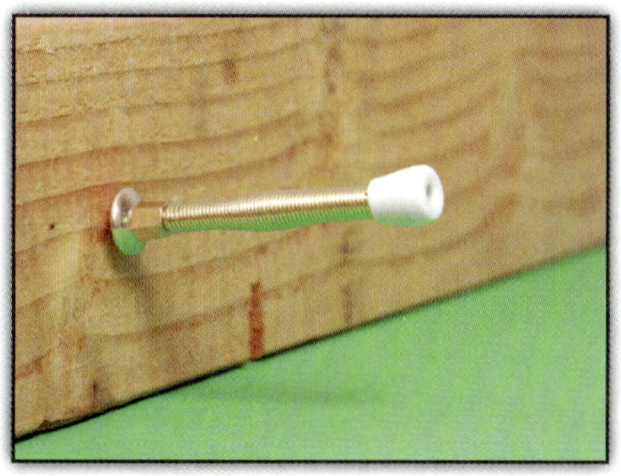

This is a doorstop. It vibrates when the door hits it.

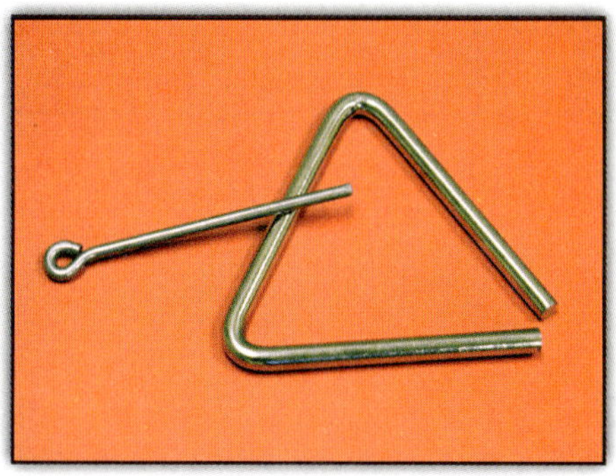

This is a triangle. The triangle vibrates when it is tapped with the striker.

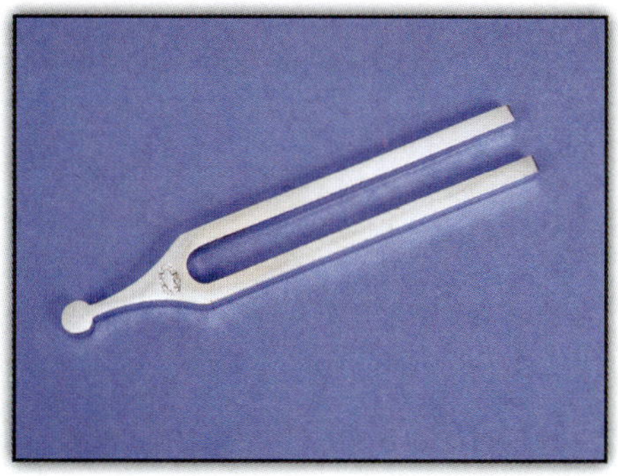

This is a tuning fork. It vibrates when it hits something.

Movement and Sound

The vibration sends a sound wave that touches the ear and helps people hear sounds.

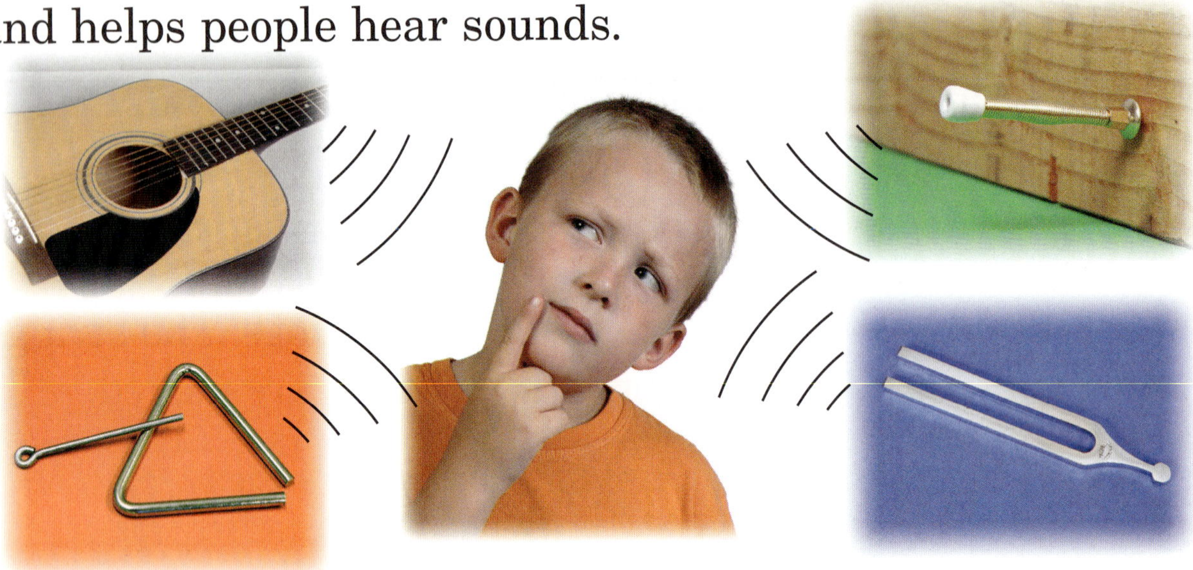

How do people hear sounds?

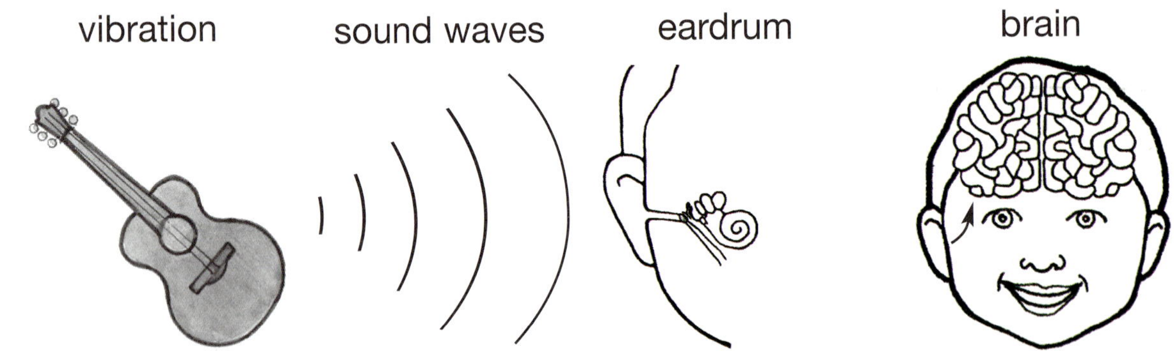

The guitar strings vibrate, sending sound waves traveling through the air. The sound waves enter the ear and make the eardrum vibrate. Nerves carry the message to the brain to tell you what sound it is.

Lesson Review

Circle the answer.

1. What word means to move back and forth very fast?

   vibrate hear sound

2. What must happen for a sound to be made?

   picture vibration habitat

Experiment

Materials

syrup, molasses, oil
3 labeled cylinders
funnels
timer

Skills

predict
observe

How does each feel?

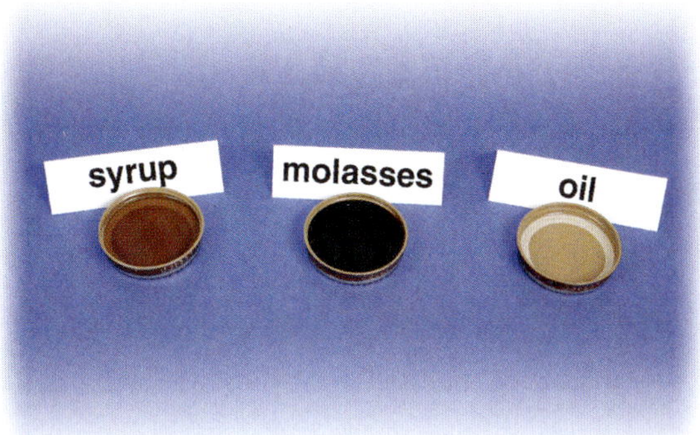

syrup ______________________

molasses ______________________

oil ______________________

Problem Which substance will move the fastest to fill the cylinder?

Predict I think ____________ will move the fastest to fill the cylinder.

I think ____________ will move the slowest to fill the cylinder.

Plan How can you test this prediction?

Experiment Use a clean funnel to pour each substance into a cylinder. Start the timer when each substance reaches the bottom of the cylinder. Watch it rise! Stop the timer when each substance fills the cylinder.

 Purposeful Design Publications

Observing Movement

Observe Draw how the syrup, molasses, and the oil look inside each cylinder. Label each cylinder.

Collect the data

Substance	Time
syrup	
molasses	
oil	

Conclusion ______________ moved the fastest to fill the cylinder.

______________ moved the slowest to fill the cylinder because it is the thickest.

Was your prediction right?

Now you know what "slow as molasses" means.

Fill in the circle below the picture if it shows movement.

○

○

○

Fill in the circle of the word that names the movement.

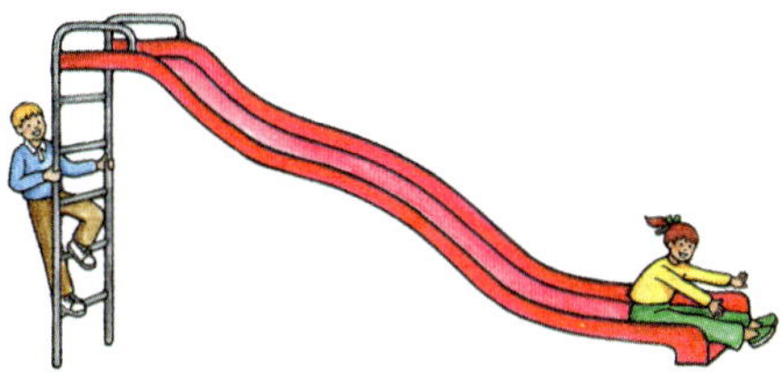

○ sliding
○ eating
○ sleeping

○ throwing
○ swimming
○ jumping

Fill in the circle. How does the merry-go-round move?

○ zigzag
○ straight
○ round and round

Write *yes* or *no*.

_______ **1.** Movement is all around us.

_______ **2.** Nonliving things can start to move on their own.

_______ **3.** A push or pull can move things.

_______ **4.** When something vibrates, it sends sound waves through the air.

Chapter 5 Review

Use the Word Bank to complete the puzzle.

Word Bank

nerves
force
substance
vibrates

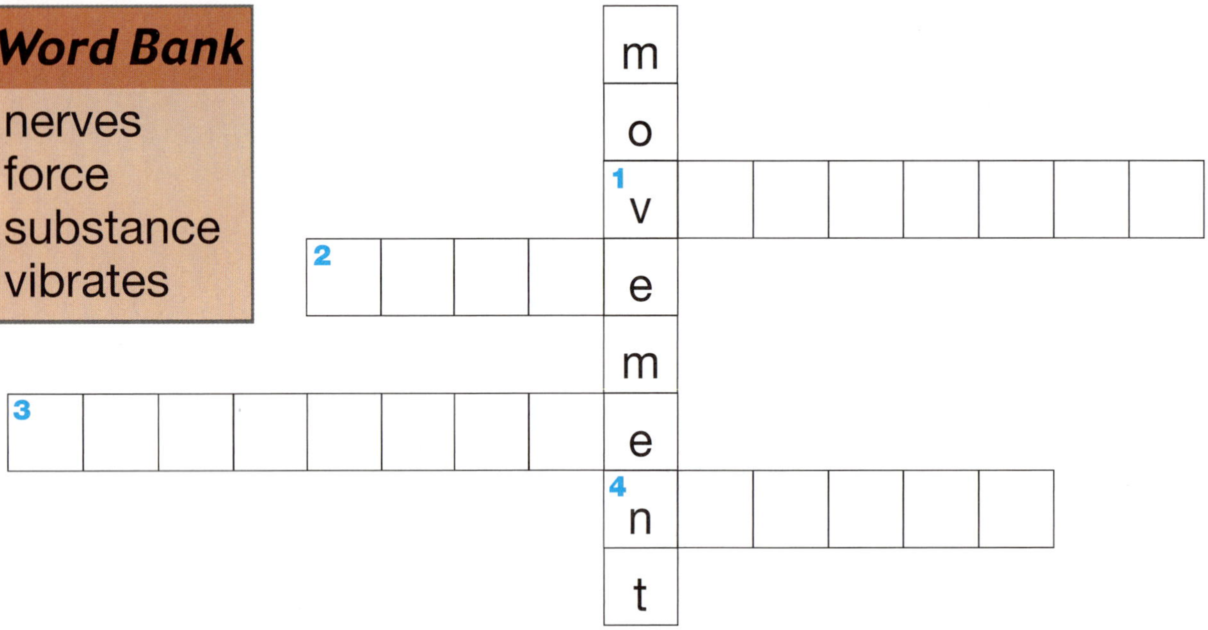

1. When something moves back and forth quickly, it ________.
2. A push is a ________.
3. Molasses is a slow-moving ________.
4. ________ carry the message to the brain to tell you what sound it is.

Journal about Movement

Machines help us.

My Booklet about Machines

Name ______________________________

Color the machine that is helping.

God gave people the ability to invent machines. Machines make work easier.

This snowblower is a machine. It helps people by blowing snow off the roads.

How does each of these things help people?

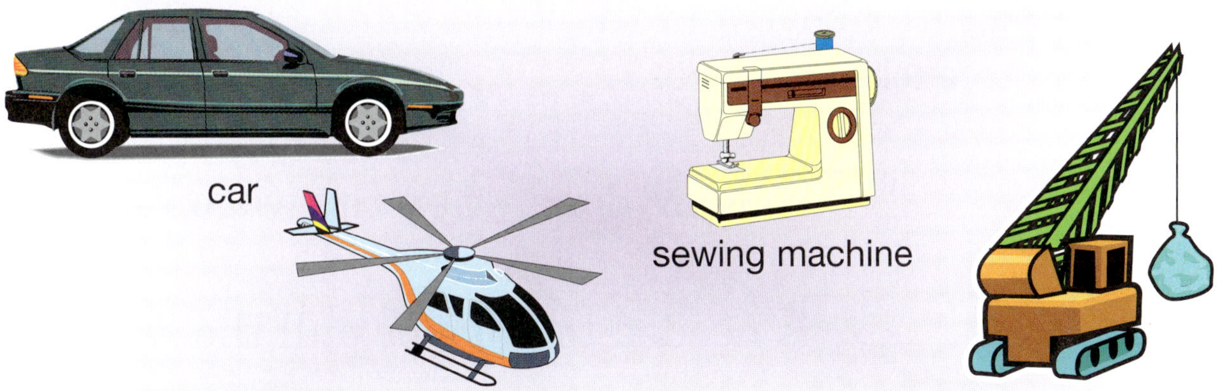

Work is done by both of these fans. The girl uses energy to provide a force that moves the paper fan. Electricity is the energy that provides the force to move the blades of the electric fan. Force is pushing the air.

Circle the fan that uses less of the girl's energy.

Machines Help Us

What are work, energy, and force?

Work is a force moving an object.

Energy is the ability to do work. **Force** is a push or pull.

Word Bank
Work
Energy
Force

Use the picture above and the Word Bank to complete each sentence.

________________ is being done when the wagon moves.

________________ is needed to push or pull the wagon.

________________ is the pull that moves the wagon.

Work is easier for people when they use less force and energy. The wheels of the wagon help the child to use less force to complete a task.

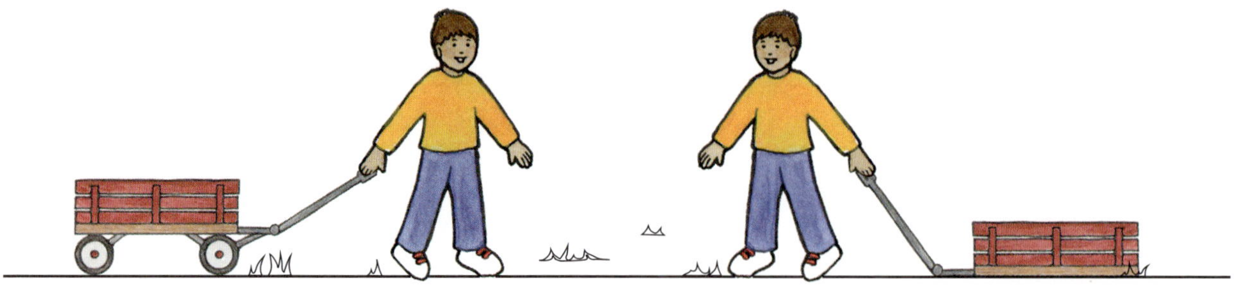

Circle the wagon that is easier to pull. What makes the task easier?

The wheel helps to move things.

Find the wheels on the cart.

What task do these wheels make easier?

Who uses the wheel?

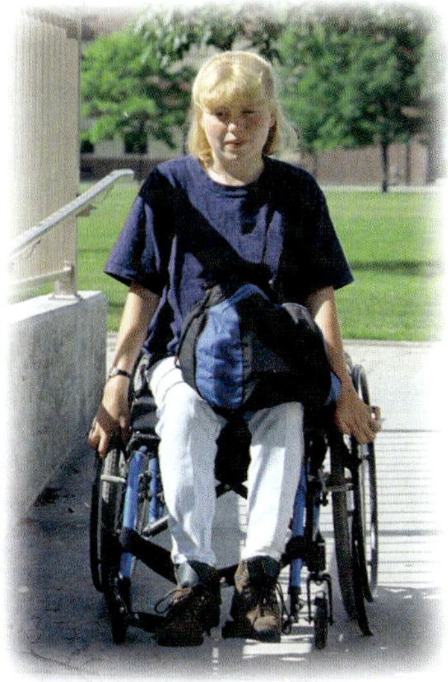

Draw a line from the picture of the wheel to the task it makes easier.

to pedal from place to place

to carry many packages of food

to go on vacation

Wheel and Inclined Plane

The inclined plane or ramp is a machine. It helps make it easier to move things. Find the inclined plane.

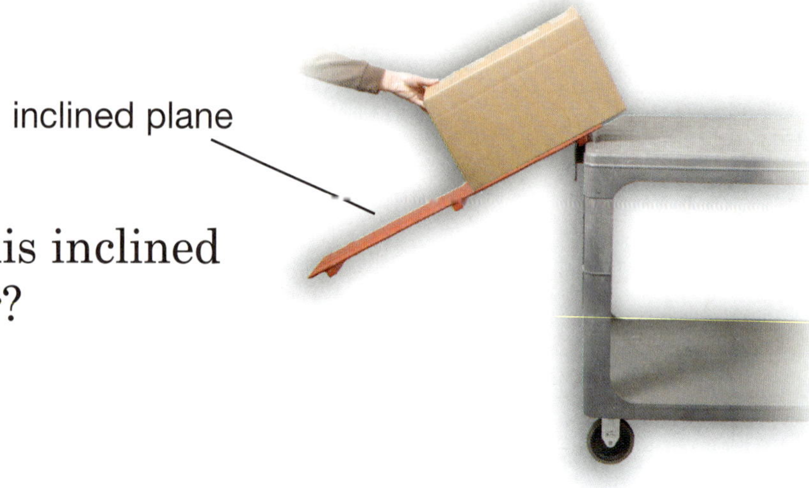

What task does this inclined plane make easier?

How is the inclined plane used?

Use the Word Bank to complete the sentences.

Word Bank	car truck

The inclined plane is part of the ______________________.

The ______________________ is moved onto the truck with the help of the inclined plane.

The lever is a machine that makes work easier.
It helps people to lift things.

Find the lever.

What task does this lever make easier?

Draw a line from each picture to the answer. What is lifted in each picture?

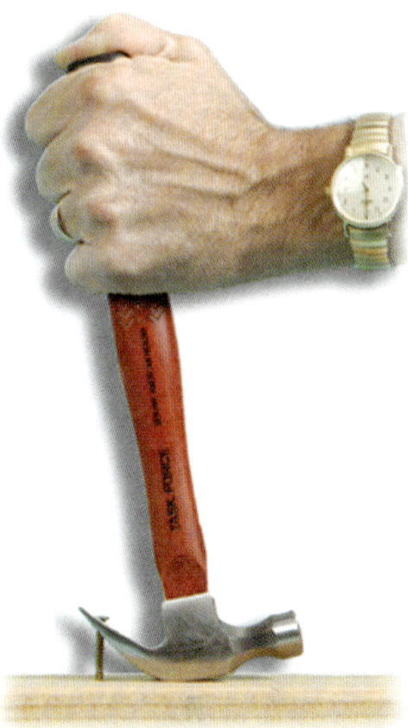

a light switch   a nail   a soda pop can

The pulley is a machine that makes work easier. It moves things by lifting them.

Find the pulley.

What task does this pulley make easier?

Use the Word Bank to complete the sentences.

Word Bank	flag pulley

This ______________________
lifts a basket.

This ______________________
is raised by a person using a pulley.

Circle the name of the machine that is doing the work.

1.  lever pulley inclined plane

2.  pulley inclined plane lever

3.  inclined plane lever pulley

Mark an *X* on the picture that does not belong.

4.   

5. 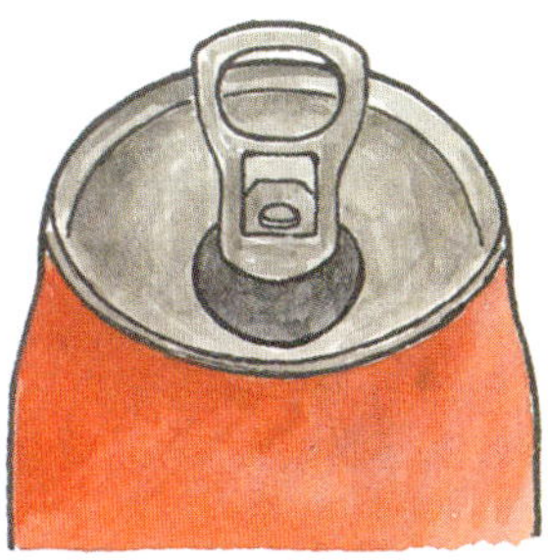  

Opening the box is a task. A lever is a machine that makes opening the box easier.

Circle the word that completes each sentence.

The force that moves the lever is a

______________________.

push wind

When the box opens, ______________________ is being done.

nothing work

______________________ is needed to push the lever.

Energy Water

Journal about a Machine

Write one way the lever and pulley are similar.

People change as they grow.

My Booklet about Teeth

Name ______________________________

Draw healthy teeth.

Some machines have gears with teeth.

A saw has teeth.

A zipper has teeth.

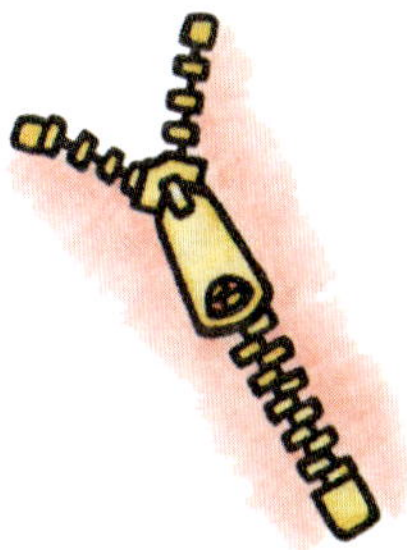

Animals have teeth.

God designed people to grow and change. One change is how teeth grow.

Most babies are born with no teeth.

Toddlers have "baby teeth."

First graders are losing their "baby teeth" and getting permanent teeth.

What are some other changes the infant makes as she grows?

I praise you because I am fearfully and wonderfully made. Psalm 139:14a

What Are Teeth?

Teeth are tools that help people to eat, smile, and talk.

Draw a line from each picture to a word that shows what teeth help people to do.

eat **smile** **talk**

Lesson Review

Word Bank
eat
God
talk
smile
taste

Use the Word Bank to complete.

1. Who designed teeth? ________________

2. Which of the five senses does your mouth help you think of? ________________

3. What do teeth help people to do?

________________ ________________ ________________

As people grow and change, their teeth do also. The first teeth are called primary teeth, or "baby teeth." By age six or seven, children begin to lose their primary teeth. Permanent teeth push on the primary teeth.

Primary Teeth

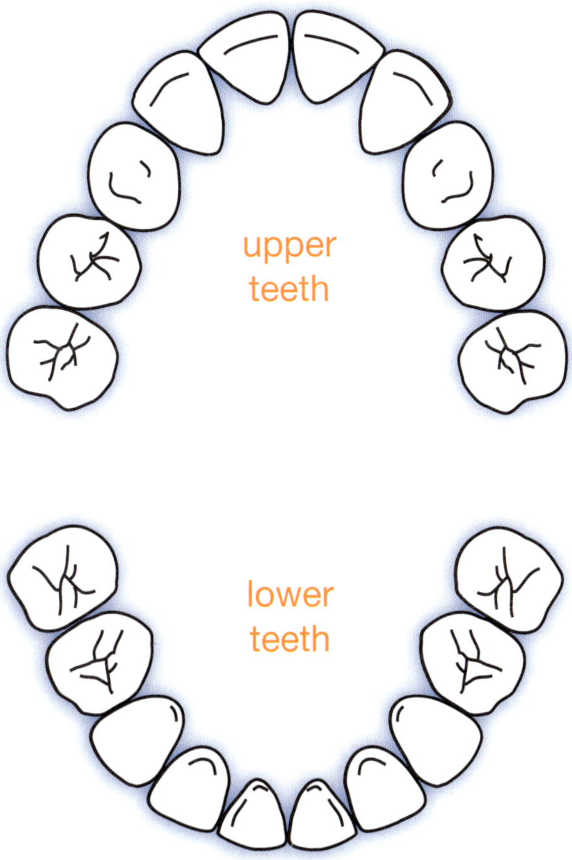

Look in a mirror and count your teeth. Write each number.

1. How many are on the top? ______________________

2. How many are on the bottom? ______________________

3. How many do you have altogether? ______________________

Use the picture of primary teeth to circle each tooth you have. Put an *X* if a tooth is missing.

Meet a tooth.

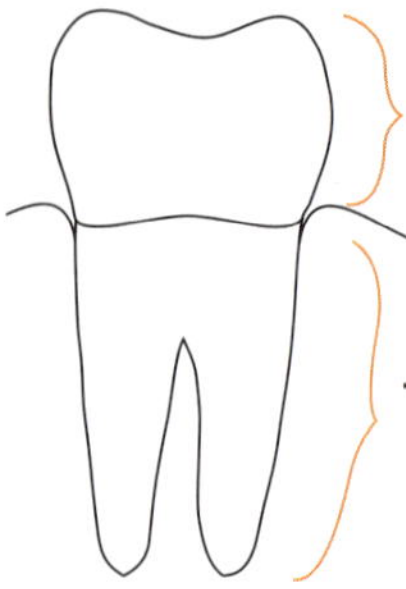

The crown is the part above the gum line that people see.

The gum is the firm tissue that helps hold the tooth in place.

The root is the hidden part under the gum line.

Write the name of each part. Use the Word Bank.

Word Bank
gum
root
crown

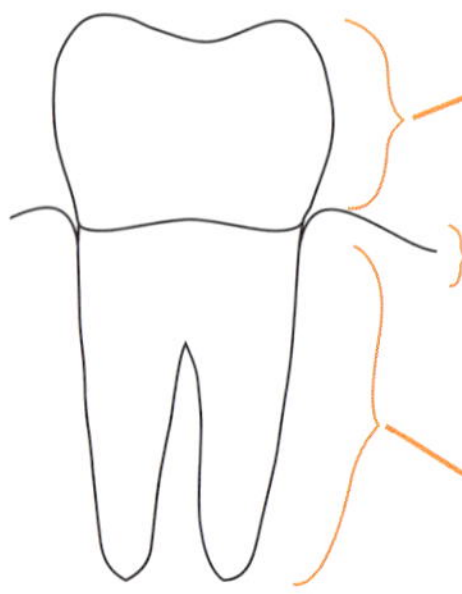

____

____

____

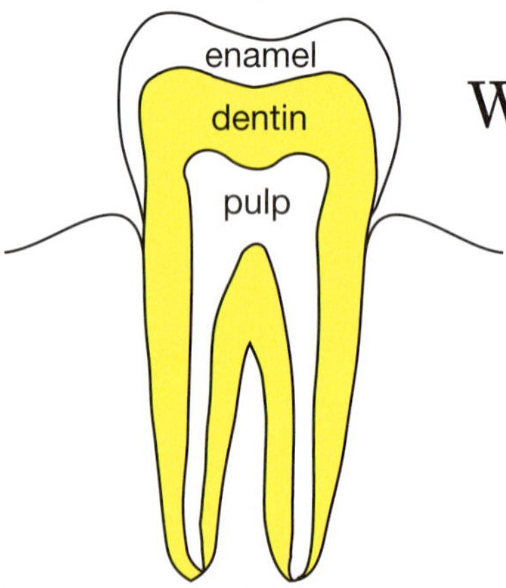

What do the parts of a tooth look like?

Leave the enamel white.
The dentin is yellow.
Color the pulp pink.

Lesson Review

Word Bank
enamel
primary

Use the Word Bank to complete.

1. Six- and seven-year-olds begin to lose their ______________________ teeth.
2. The crown of the tooth is covered with ______________________.

Do different types of teeth have different jobs?

Incisors bite and cut food.

Cuspids bite and tear food.

Molars grind and chew food.

Primary Teeth

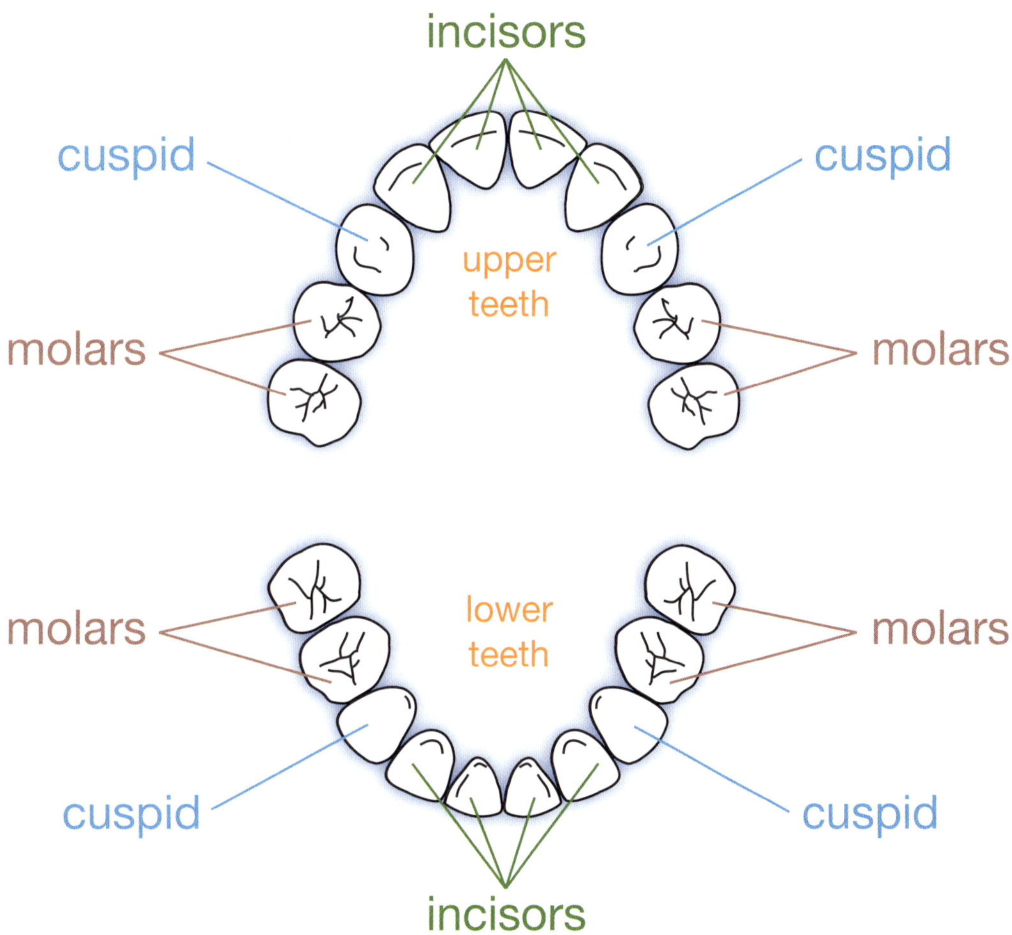

Write *yes* or *no*.

_______ **1.** Incisors appear in the upper teeth only.

_______ **2.** Cuspids appear in the lower teeth only.

_______ **3.** Incisors, cuspids, and molars appear in the upper and lower teeth.

Types of Teeth

How does the shape of the tooth fit the job of the tooth?

Incisors are sharp and in the front of the mouth, so they can bite and cut.

Cuspids have a pointed sharp cusp, so they can bite and tear.

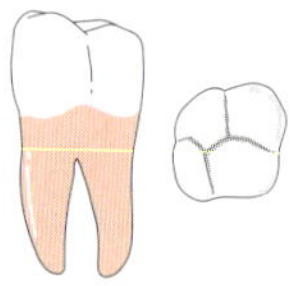

Molars have broad tops and cusps with pointed edges, so they can grind and chew.

Color.

incisors 

cuspids 

molars 

Primary Teeth

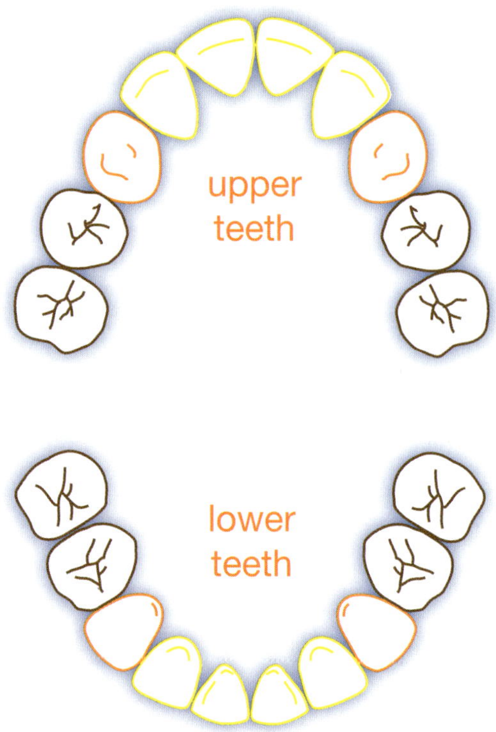

Lesson Review

Circle the words that tell about teeth.

ring	cuspid	wind	bite	eat
incisor	chew	cut	book	smile

How do teeth stay healthy?

Brush your teeth. This helps to remove plaque that causes decay.

Floss your teeth. This cleans between teeth where a brush cannot reach.

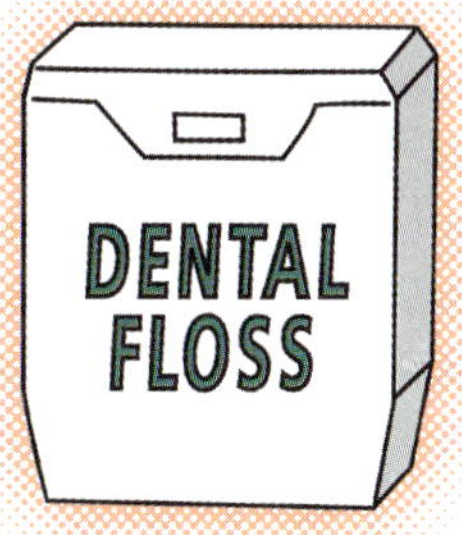

Wear protection to keep teeth safe when playing sports.

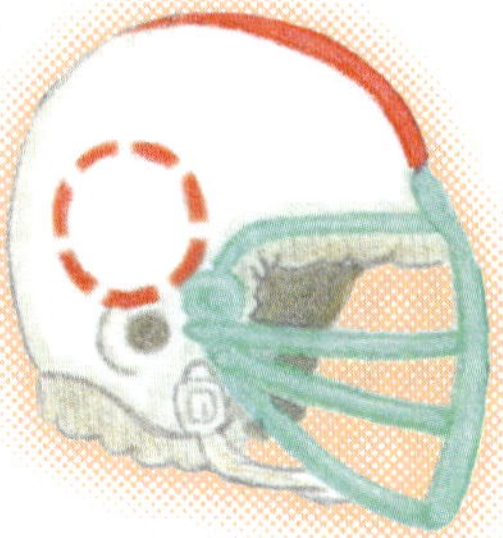

Eat healthy foods.

Circle the foods that help to keep teeth healthy.

Experiment

Materials		
hard-boiled egg	toothpaste	container
toothbrush	can of cola	

Skills
observe
experiment

Day 1

Problem Will cola stain an egg?

Observe What color is the egg?

Predict If the egg is placed in cola today, what color do you think the egg will be tomorrow?

____

Plan How can you test this prediction?

Experiment Place the egg in a container of cola.

Day 2

Collect the data What color is the egg that was placed

in cola? ______________________________

Conclusion ______________________________

Experiment further Brush the stained egg with toothpaste.

____

Conclusion ______________________________

____

Was your prediction right?

The Dental Office

What does the dentist do?

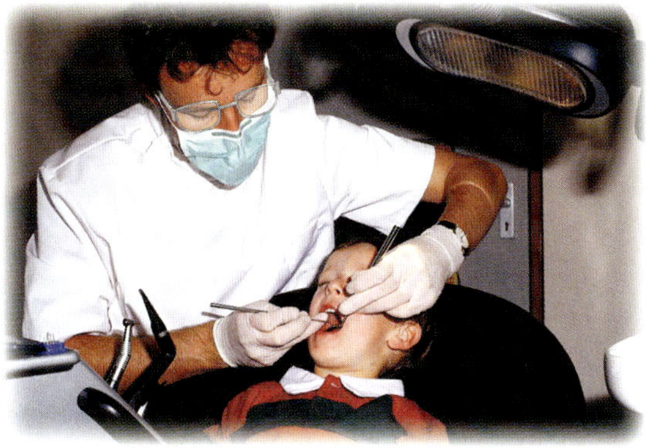

- Treats teeth, gums, and other parts of the mouth.
- Looks for teeth that have grown the wrong way.
- Counts and keeps records of each tooth.
- Checks for cavities and fills them.

What does the dentist wear?

Rubber gloves 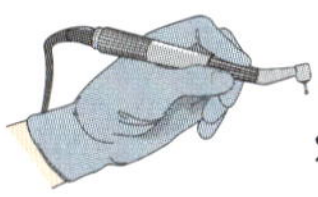, a mask 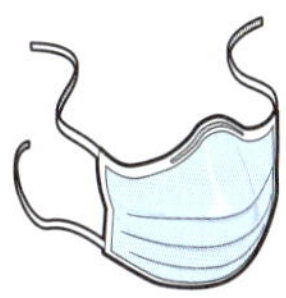, glasses , and a coat 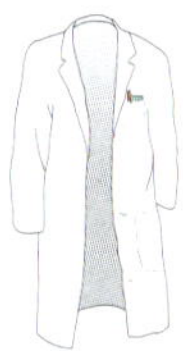 help keep germs from spreading.

What tools does the dentist use?

The explorer is used to count the teeth and check for decay.

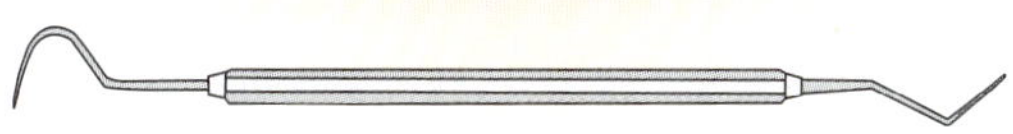

The dental mirror is used to see the back of teeth.

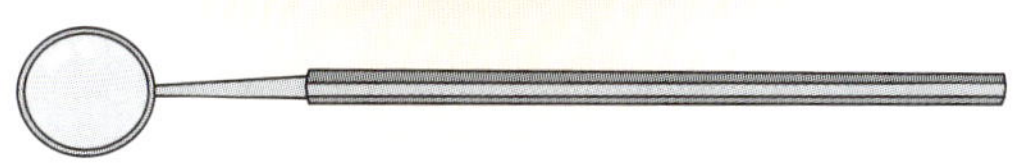

This tube removes water from the mouth.

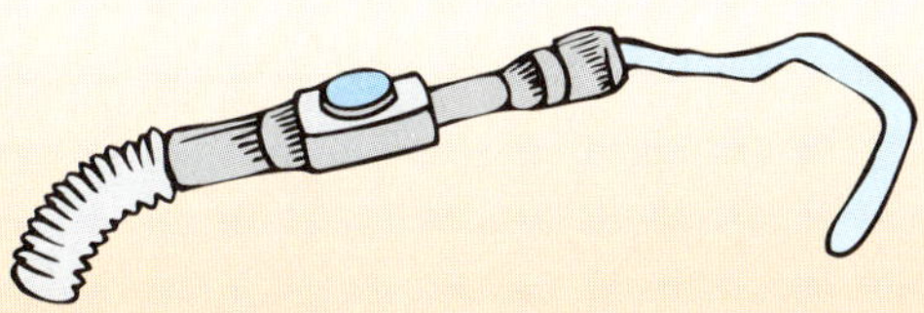

An x-ray machine takes pictures of teeth, gums, and bones.

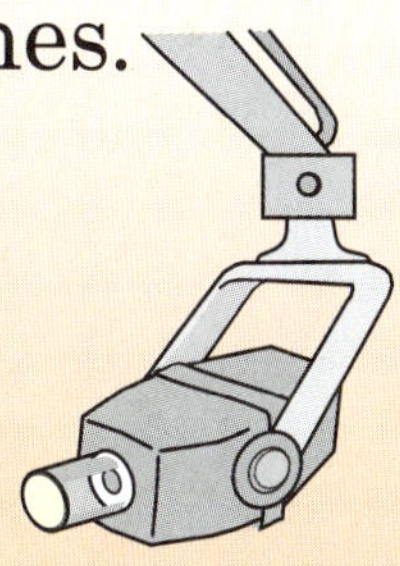

Color 15 objects hidden in the tree.

Objects

mouth with braces	dental light	apple
toothbrush	dental chair	glass of milk
toothpaste	x-ray machine	carrot
floss	tooth	broccoli
mouthwash	smile	cheese

Fill in the circle to complete each sentence.

1. One of the jobs of teeth is to _____.

   ○ think ○ eat

2. "Baby teeth" are not _____.

   ○ permanent ○ primary

3. A tooth's crown is _____ the gum line.

   ○ above ○ below

4. Enamel, dentin, and _____ are parts of a tooth.

   ○ skin ○ pulp

5. Color the teeth.

Color the incisors yellow.

Color the cuspids blue.

Color the molars brown.

Primary Teeth

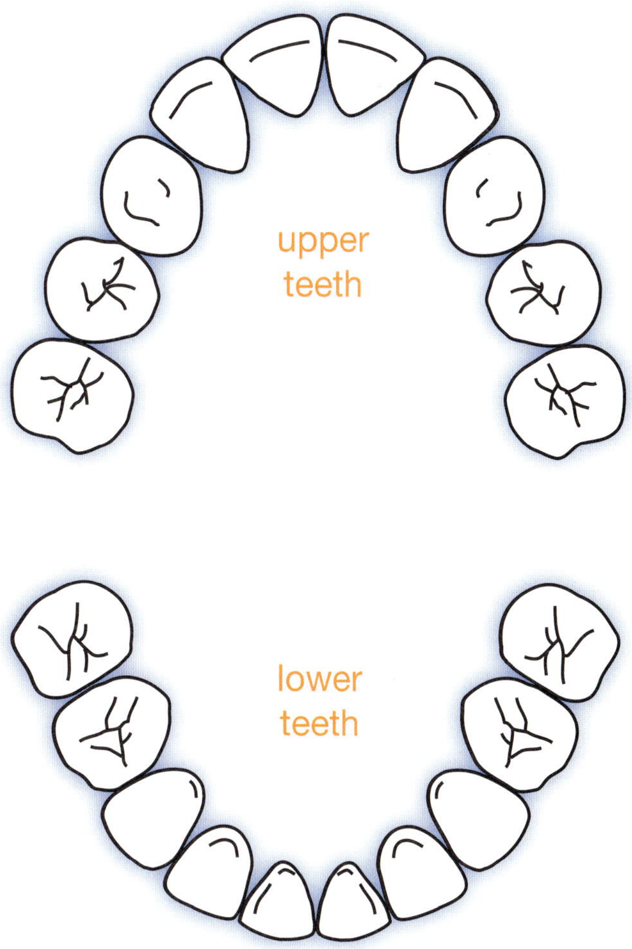

Chapter 7 Review

Read the creation web.

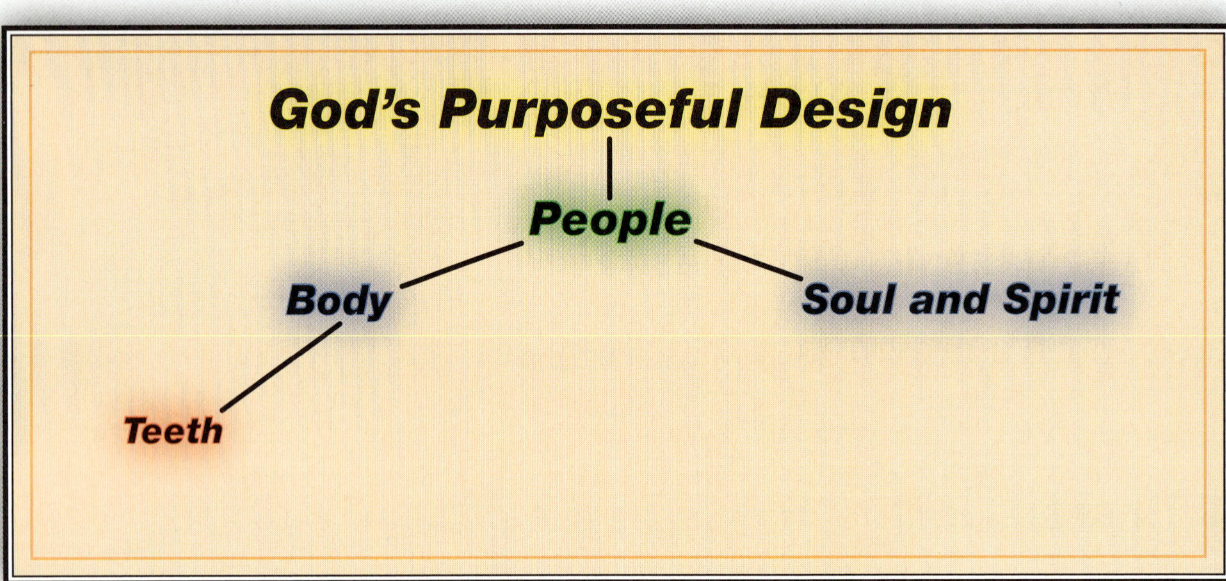

Journal about Teeth

I learned

Keep your bones and muscles healthy.

My Booklet about Bones and Muscles

Name ______________________________

"Please sign and color my cast."

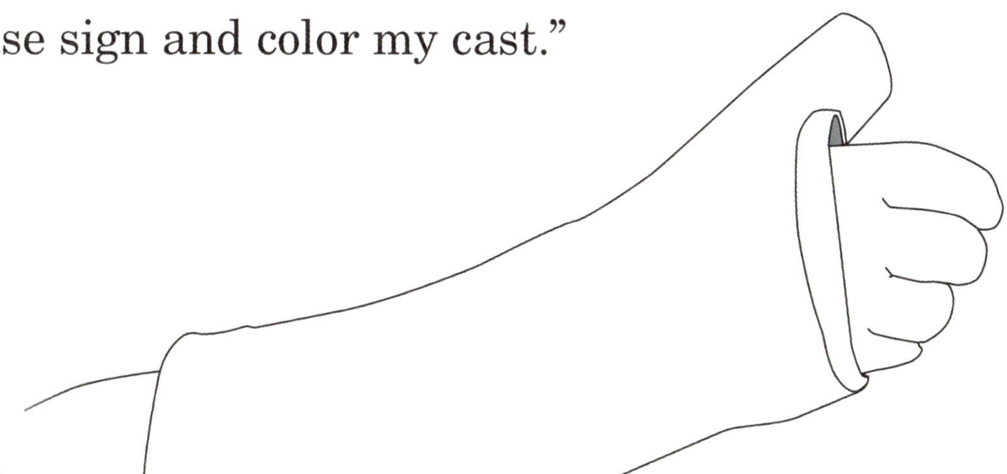

Bones are a hard part of the body. Bones give the body a shape and help it move.

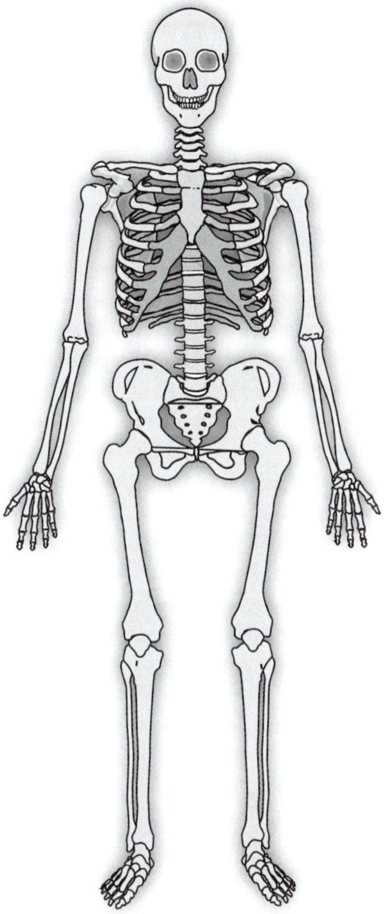

Where are this child's bones?

Bones grow inside the body. You cannot see them, but you can feel some of them under your skin. All your bones make up a skeleton. Use your finger to trace the child's skeleton.

God designed the bones in your body to work together to help you move and to give you a shape.

Your hands shaped me and made me....
and knit me together with bones. Job 10:8a, 11b

What Are Bones?

Use your finger to trace each skeleton below.

Is the skeleton of a mammal, a fish, a bird, and an insect on the inside or outside of the body? Circle your answer.

inside or outside

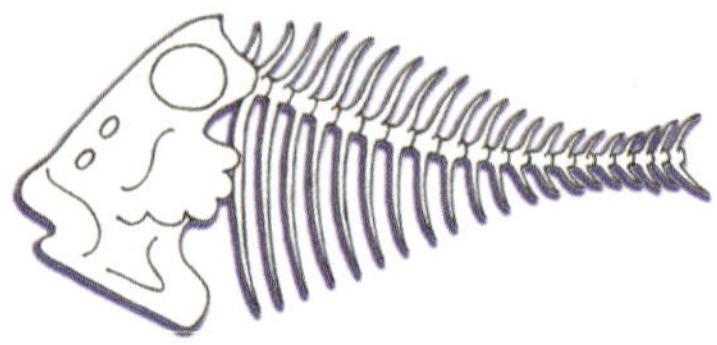

inside or outside

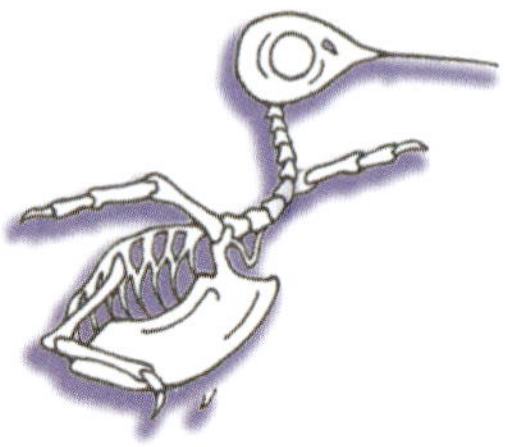

inside or outside

inside or outside

Observe the skeleton in each picture above.
Do all the animals have the same shape? Explain.

What type of animal do you think this skeleton is from?

Lesson Review

Use the Word Bank to complete each sentence.

Word Bank
move
shape

1. Bones give the body a ________________.

2. Bones help the body to ________________.

Why do football players wear helmets?
Is a helmet soft or hard?

Blow up a balloon.
Is a balloon soft or hard?

Place the balloon inside the helmet.
Is the balloon more protected inside the helmet or outside the helmet?

The helmet protects the balloon like the skull protects the brain.

The skull, ribs, and backbone are bones.

Feel your skull. How does it feel?
The skull protects the brain.

Feel your ribs. How do they feel?
The ribs protect the heart and lungs.

Feel your backbone. How does it feel?
The backbone protects the spinal cord.

Bones are hard. They protect the soft parts inside the body.

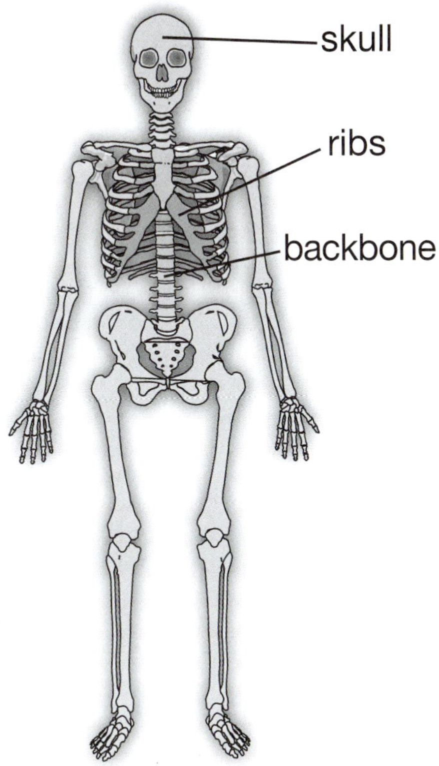

Know that the Lord is God.
It is he who made us. Psalm 100:3a

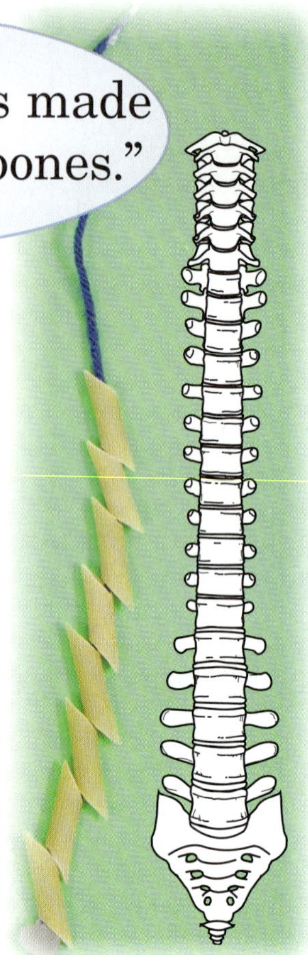

What does the backbone protect?

Lesson Review

Write what you know about bones.

More about Bones

How many bones are there in the body?

Make a guess. ______________________ bones

Observe the bones of the skeleton.
Observe the bones around the skeleton.

Do you think the number of bones in the skeleton is the same as the number of bones around the skeleton?

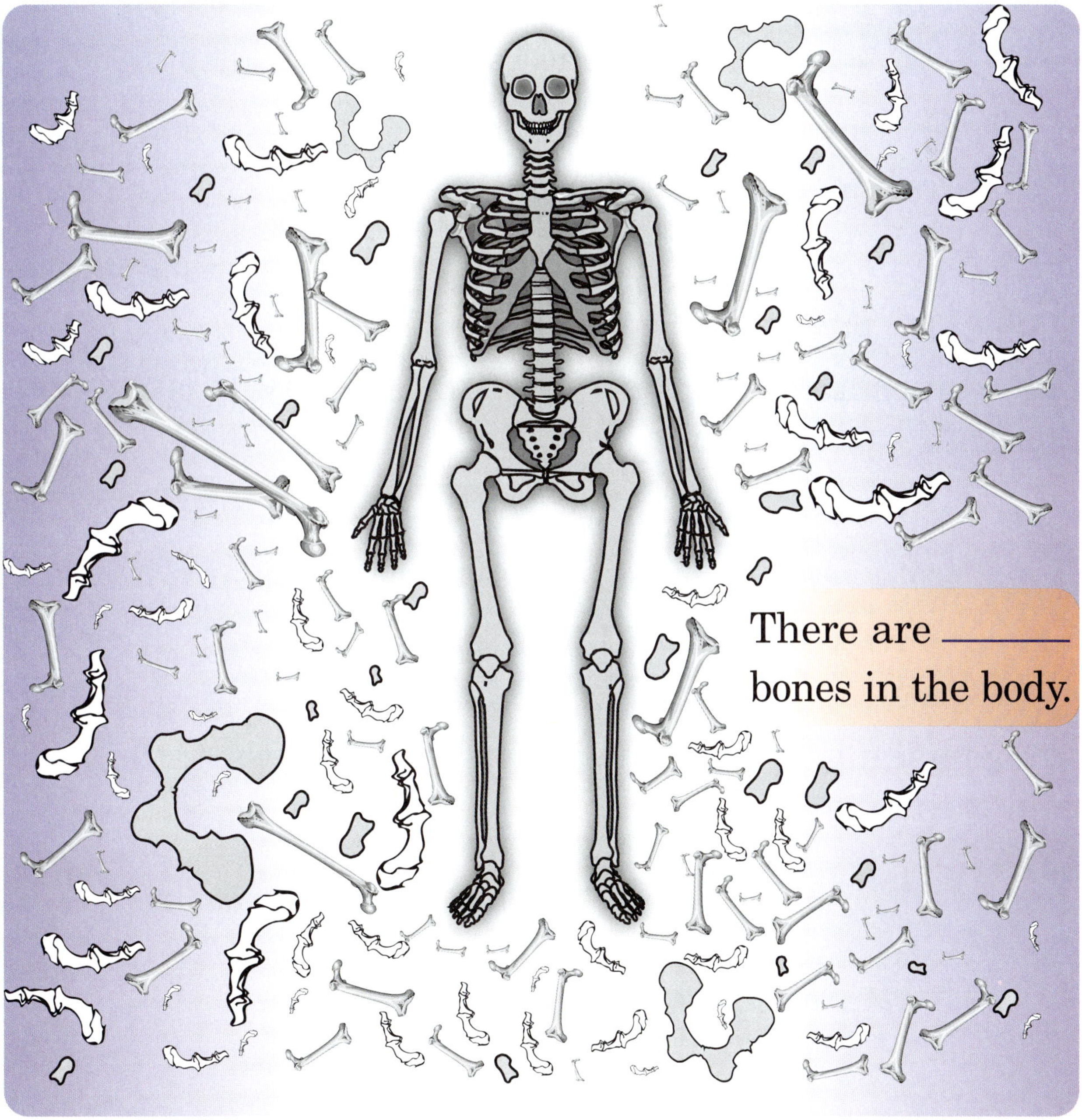

There are ______ bones in the body.

More about Bones

What is the smallest bone in the body?
What is the largest bone in the body?

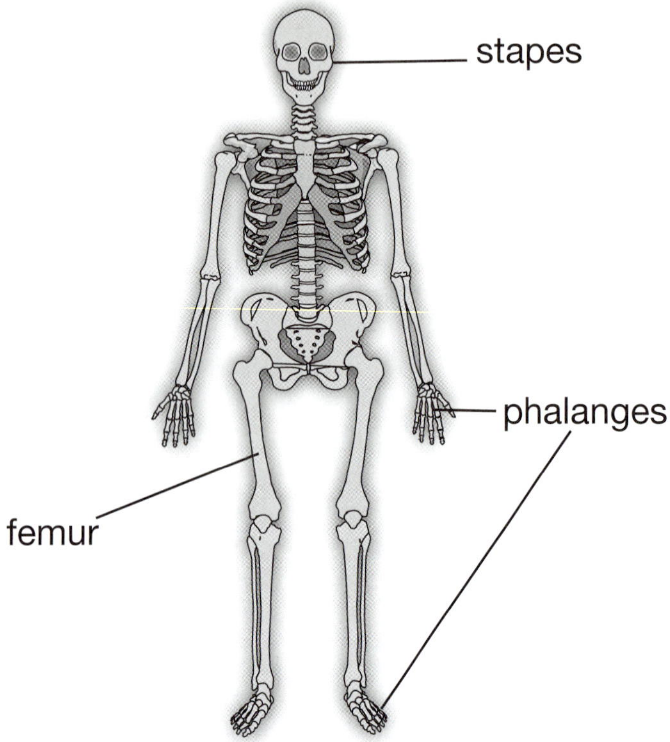

A stapes looks like this.

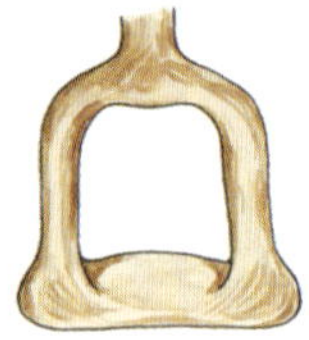

It is really only this big.

The stapes is the smallest bone in the body. It is inside the ear. It helps people hear.

The femur is the largest bone in the body.
Touch your femur. Use an inch ruler to measure it.

How many inches long is your femur? ____________ inches

Your phalanges are the bones in your fingers, thumbs, and toes.

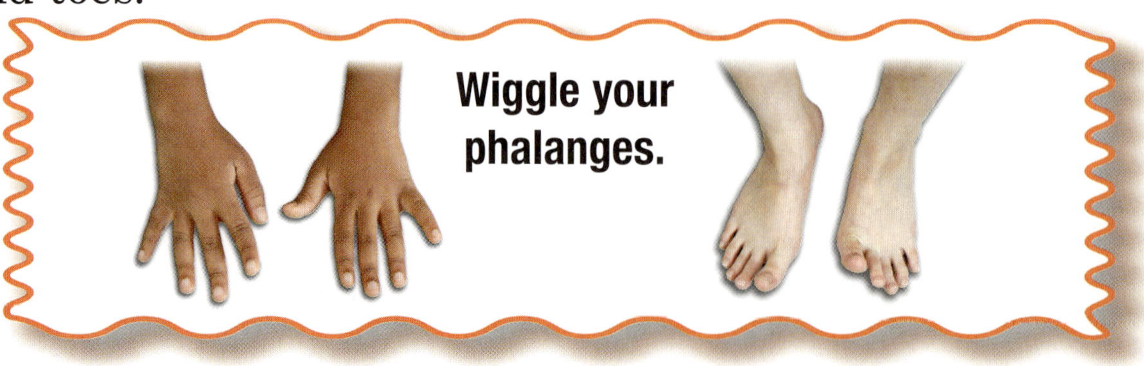

What makes healthy bones?

Certain foods help bones grow and stay strong.

There are groups of food that you should eat each day.

Foods from the milk, yogurt, and cheese group have calcium in them. It is important to eat many foods with calcium to grow strong bones.

Exercise helps bones grow and stay strong. When you move your body, you are exercising.

How are these children exercising?

How do you exercise?

Healthy Bones

What happens when a bone breaks?

This child fell off her bike.
A doctor looks at her painful arm.

X-rays are taken of the arm.

The x-rays show a small crack in the bones.

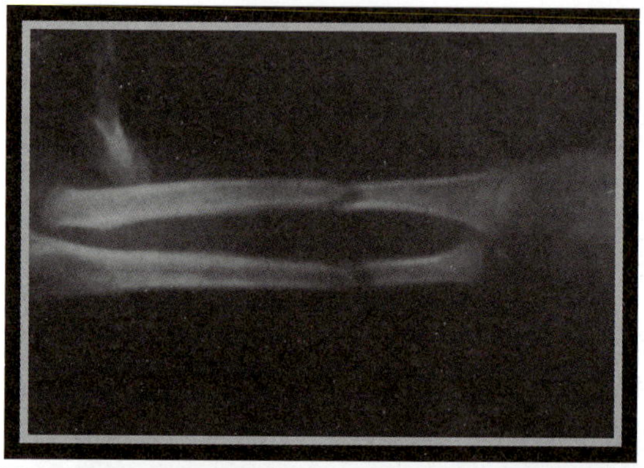

The doctor sets the bones.

A cast is put on the arm until the bones heal.

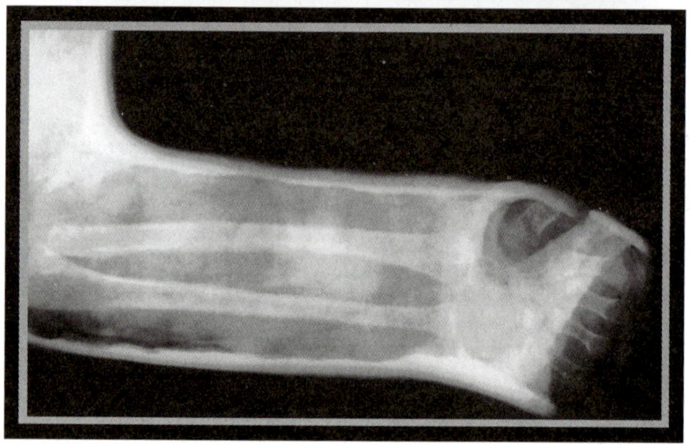

Lesson Review

Write *yes* or *no*.

__________ **1.** Certain foods help bones grow strong.

__________ **2.** Calcium is good for your bones.

__________ **3.** You should not move your bones.

Muscles move parts of your body by moving your bones.

Flex your arm muscle.
Can you feel that muscle?

Place your finger on an arm muscle in the picture.

A **joint** is the place where bones meet. Bones cannot bend. Joints help you bend, twist, or turn.

What helps these children to bend, twist, or turn?

Muscles and Joints

Hinge joints help you bend your knees and elbows.

A hinge joint works like the hinge on a door.

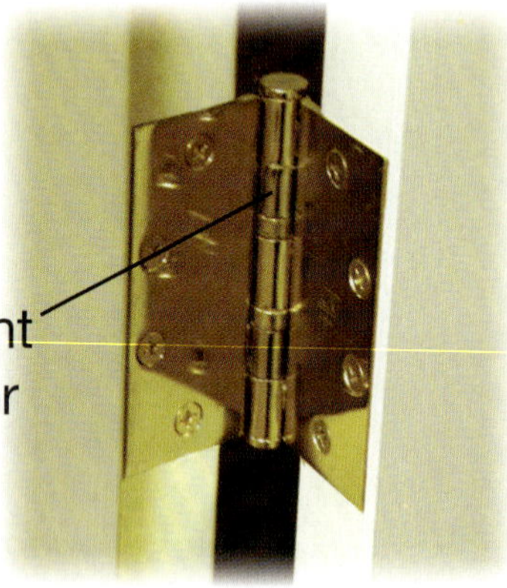

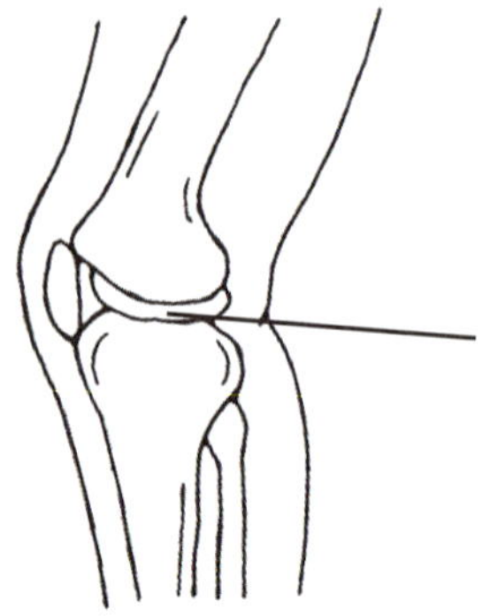

Why do your legs and arms stop bending when they are straight out?

Ball and socket joints help you move your whole arm and whole leg.

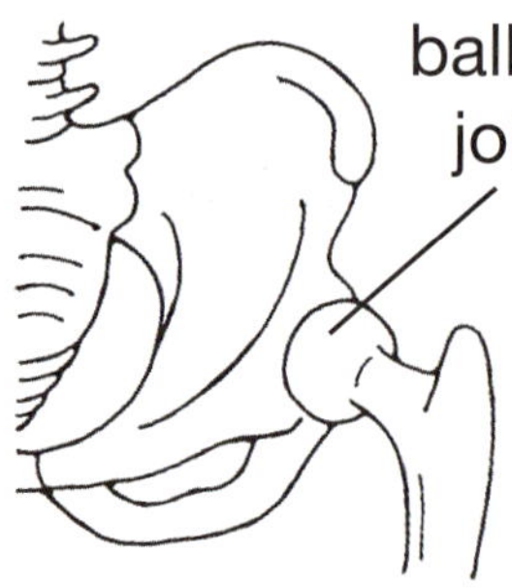

O Lord, you are our Father. We are the clay, you are the potter; we are all the work of your hand. Isaiah 64:8

Lesson Review

Circle the word that completes each sentence.

1. __________ help bones move.

   Apples Muscles Games

2. The place where __________ meet is a joint.

   teeth boxes bones

Fill in the circle that answers each question.

1. What is the name for all the bones in your body?

   ○ muscle ○ skeleton ○ joint

2. What helps your bones move?

   ○ muscles ○ houses ○ birds

3. What is one job of the skeleton?

   ○ to plant trees
   ○ to feed animals
   ○ to protect the soft inside parts of the body

Identify each bone. Draw a line from the word to the bone on the skeleton.

skull

ribs

backbone

femur

phalanges

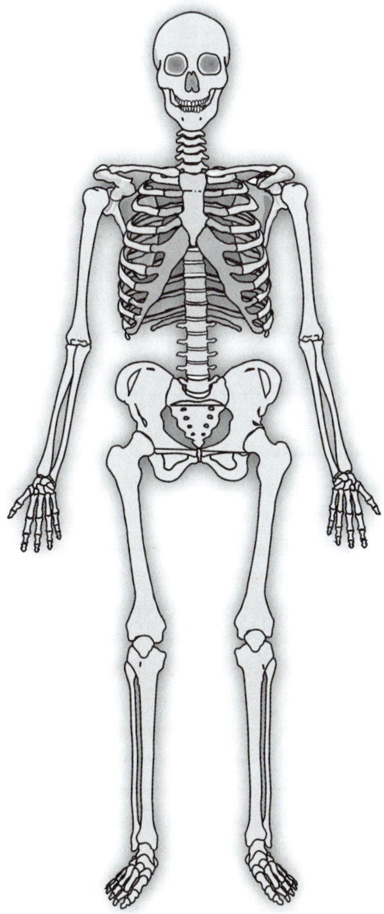

Read the creation web.

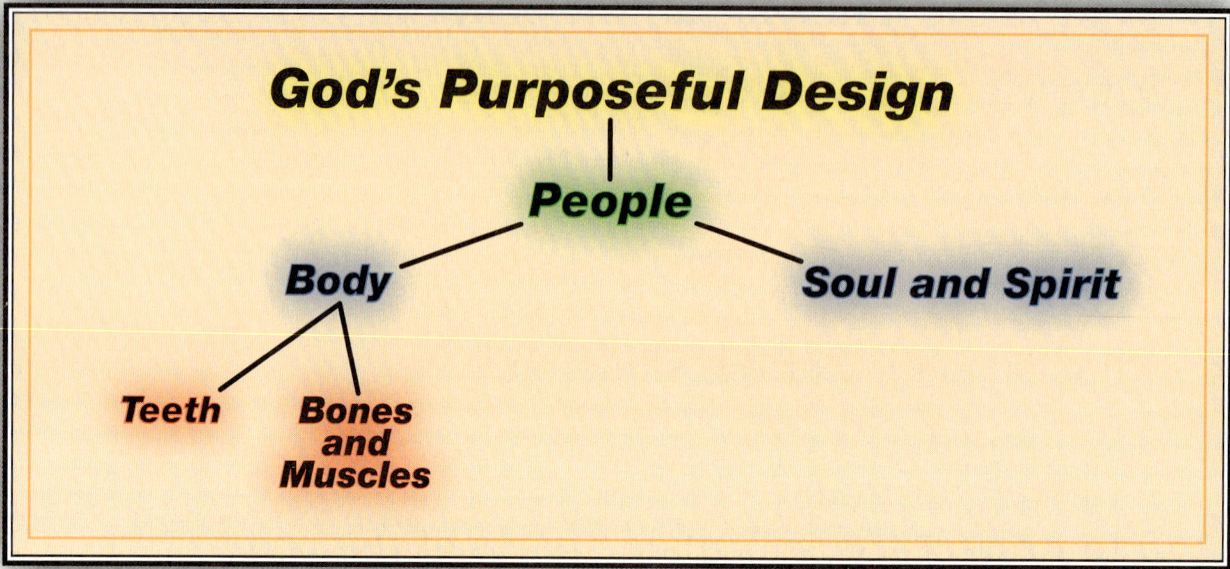

Journal about Bones and Muscles

Write what amazed you about bones and muscles.

Exercise keeps your heart healthy.

My Booklet about the Heart and Blood

Name ________________________________

Draw someone exercising.

What Is the Heart?

The heart is an organ that pumps blood to all parts of the body.

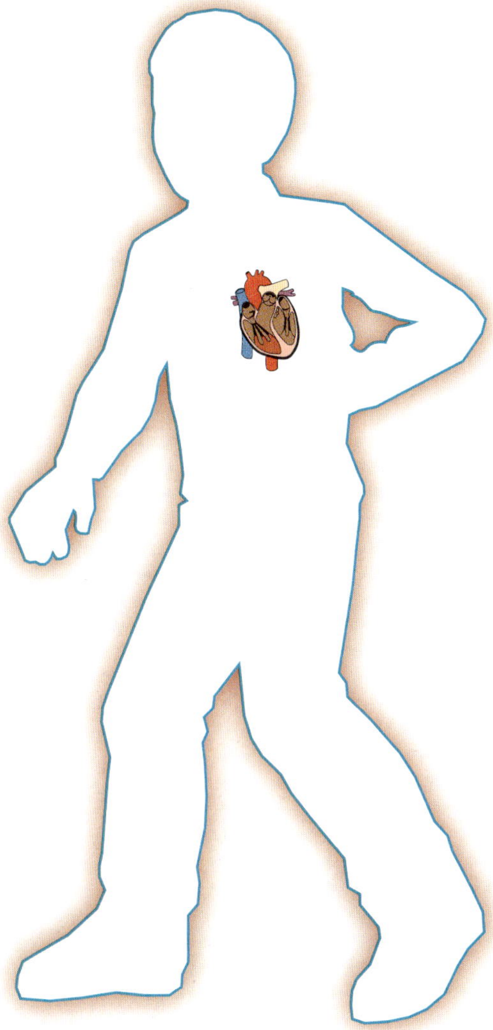

Draw a picture of your heart.

Can you feel your heart beating? Place your hand over your heart.

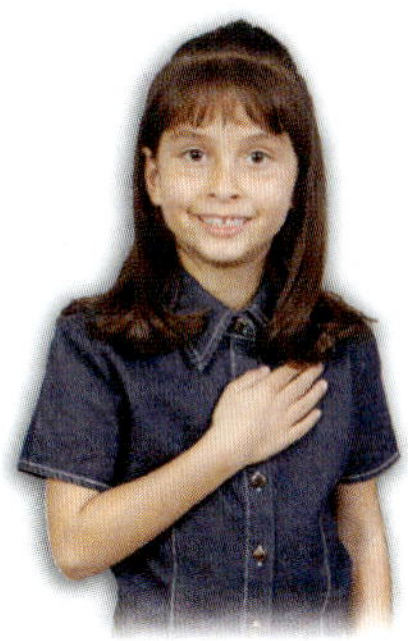

What are some ways to check your heart rate?

Feel the side of your neck.

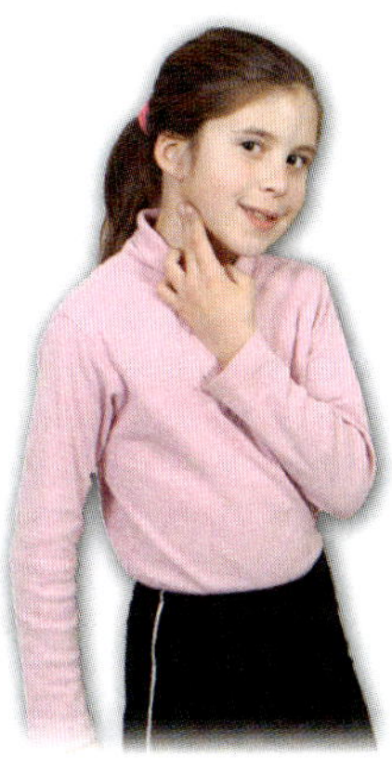

Feel your arm.

Blood keeps moving through the body because it is pushed by a strong pump called the heart.

What Is the Heart?

Can you hear your heart beating?

A doctor uses a stethoscope to listen to the child's heart.

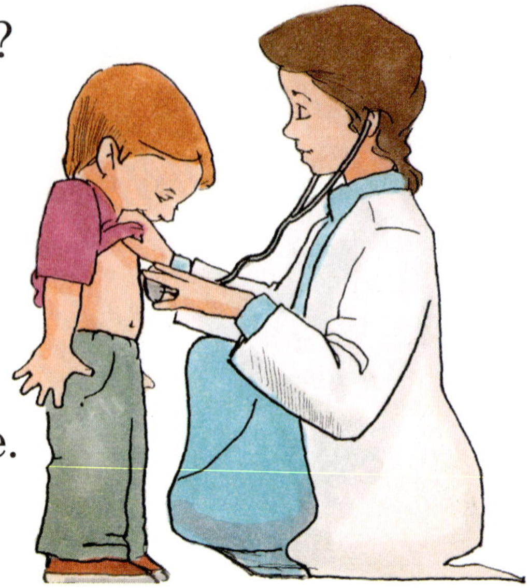

Make a stethoscope by using a paper cup and a paper towel tube. Use it to listen to the beating of someone's heart.

Lesson Review

Circle the picture that each sentence tells about.

1. It pumps blood to parts of the body.

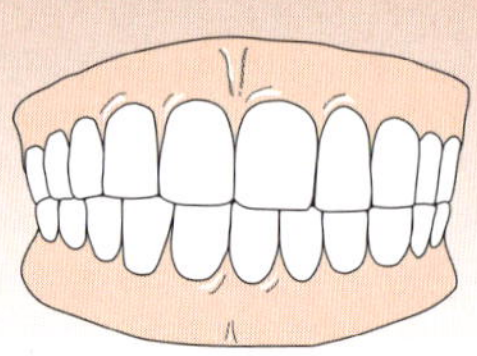
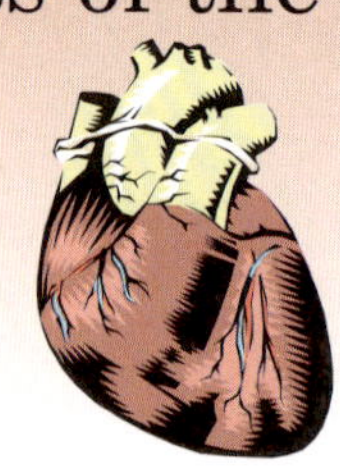
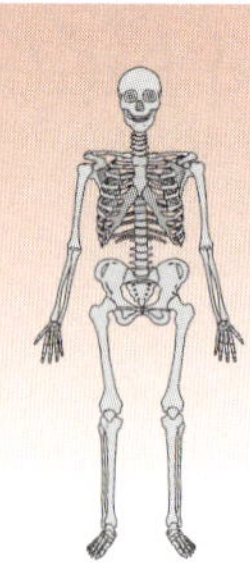

2. It helps someone hear the beat of the heart.

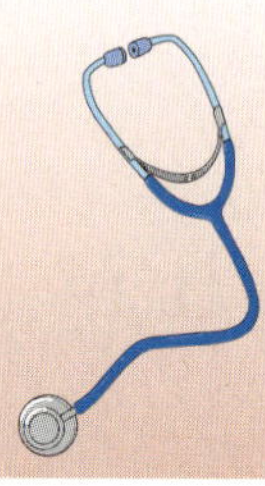

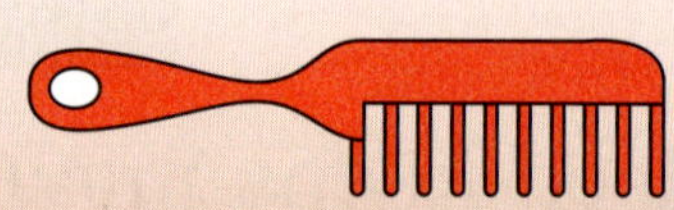

What is the heart made of?

The heart is a muscle.

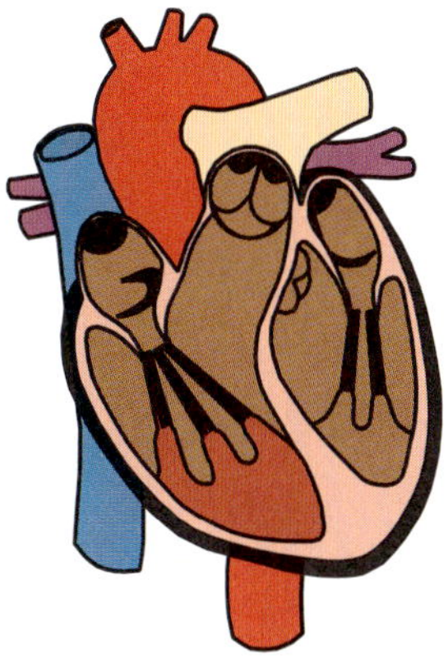

How big is the heart?

It is about the size of your fist.

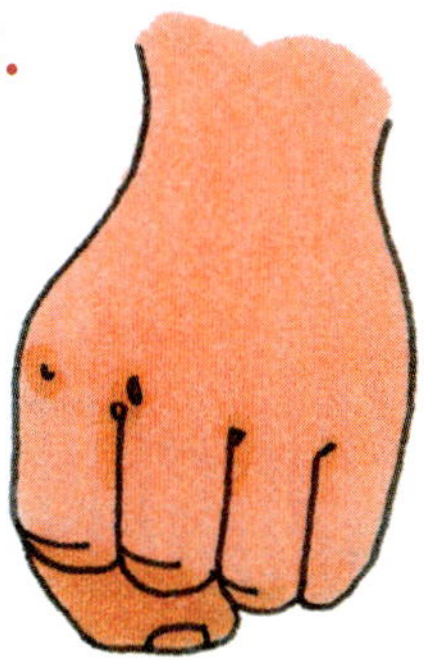

How much does the heart weigh?

It weighs less than one pound.

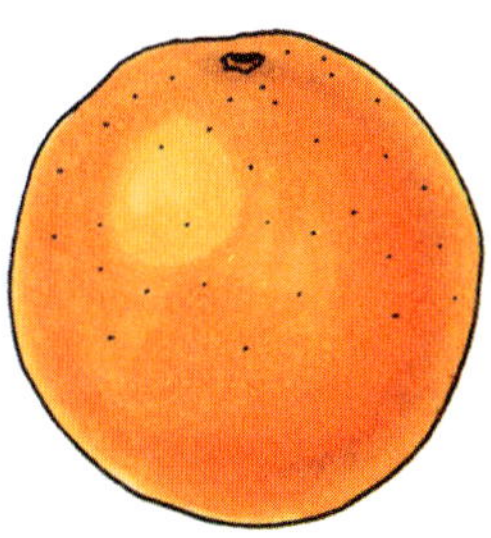

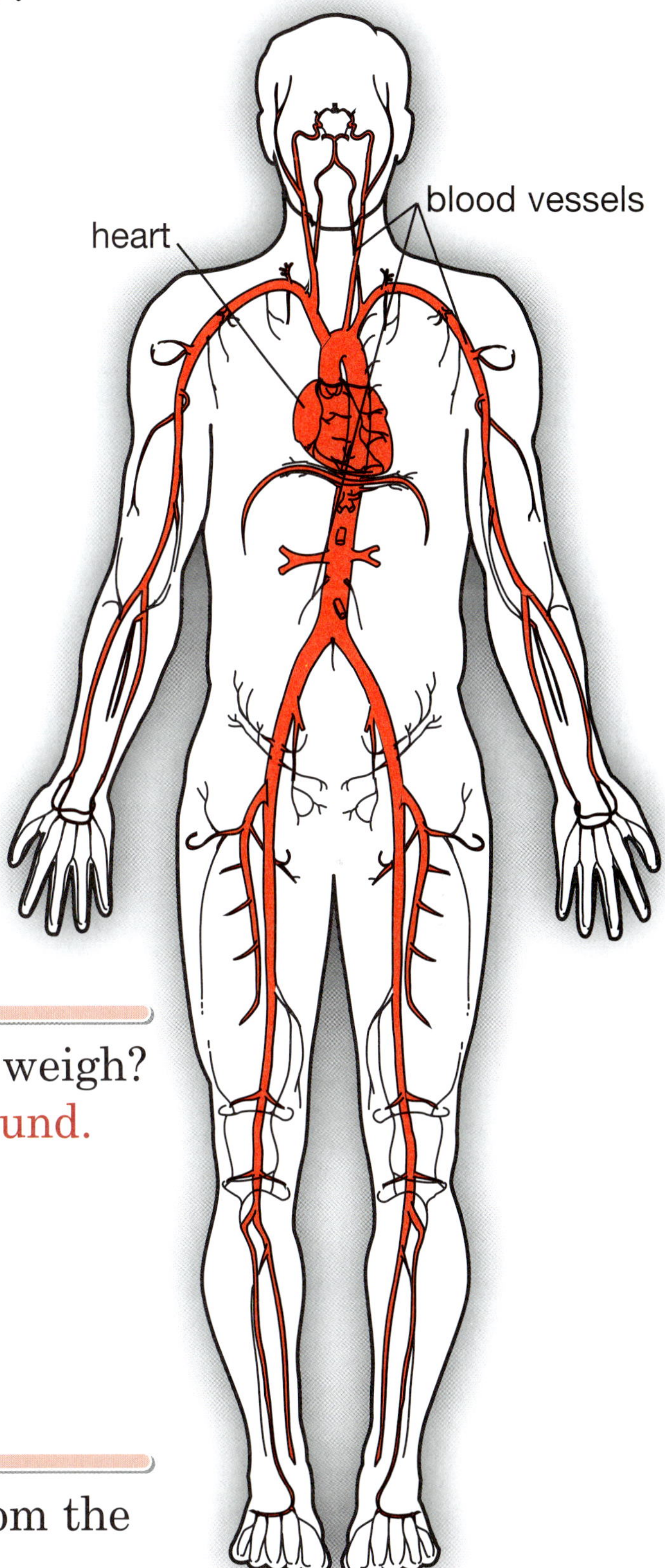

How does the blood get from the heart to the toe?

It flows through blood vessels. It carries the things we need to stay alive. Trace the blood vessels with a finger.

More about the Heart

Explore how the heart pumps blood through blood vessels. You will need a cup of water, a dropper, and straws.

Follow the steps.

1. Fill the dropper with water.
2. Place the tip of the dropper inside the top straw.
3. Pump or squeeze the dropper.
4. Observe the flow of water.

What makes the blood in your body flow?

Draw a picture of your heart and blood vessels.

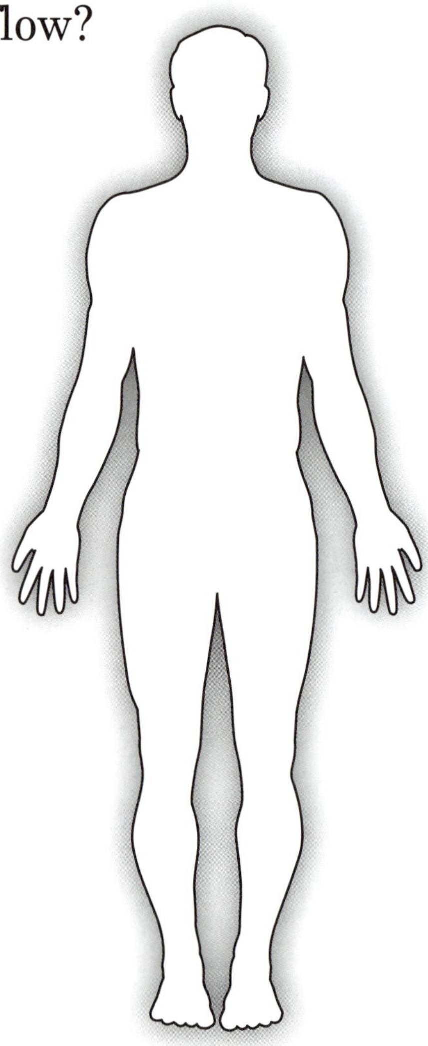

Cuts Heal

How does a cut heal?

1. The skin on the arm is broken. The cut bleeds.

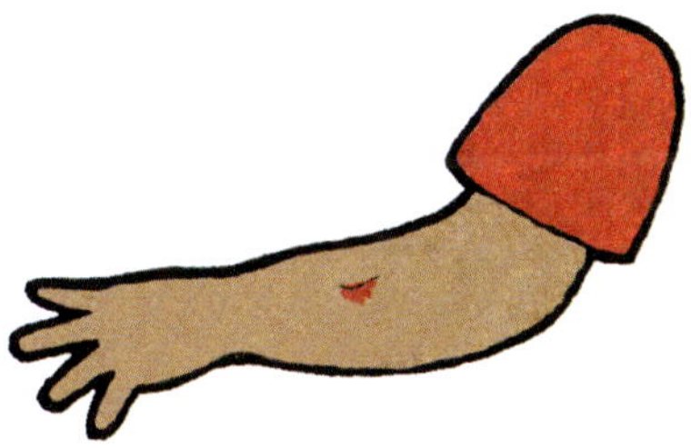

2. The cut is washed with water and soap. A bandage is put on the cut.

3. The body has its own way to stop bleeding. It forms a scab.

4. New skin forms under the scab. The scab falls off.

Design and color a bandage. It can be the shape of a rectangle 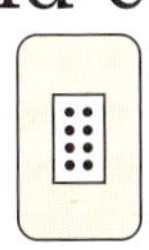, square , or circle 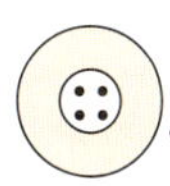.

Cut it out and use it on the next page.

Cuts Heal

Draw someone with a small cut. Use the bandage you made to cover the cut. Glue.

Lesson Review

Write 1 through 4 on the blanks to show how a cut heals.

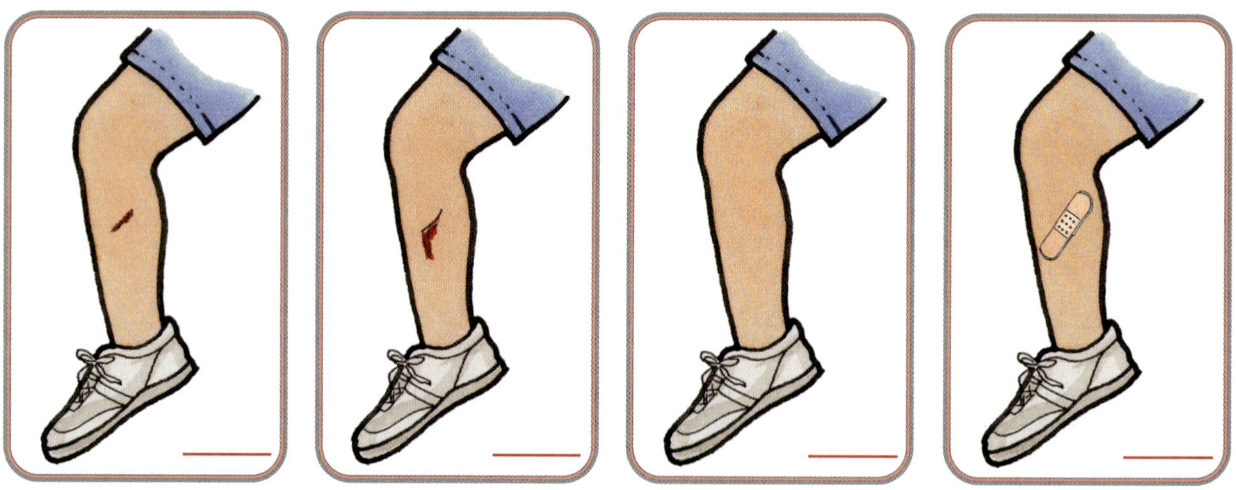

Tech Connect

What makes a healthy heart?
A healthy heart needs exercise.

A healthy heart needs food such as fruit, vegetables, milk, bread, and meat with little fat.

Try this.
Place some potato chips on a piece of colored paper. Wait an hour. What do you see on the paper?

Too much fat is not good for your heart.

Which road leads to a healthy heart? Color it.

Tech Connect

God has given doctors the ability to help people take care of their hearts.

An electrocardiogram is a test that checks out a person's heart. A healthy heart recording looks like this.

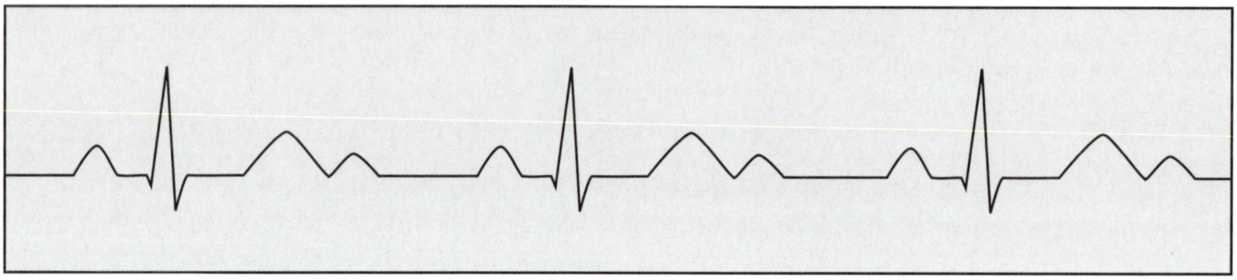

An unhealthy heart recording can look like this.

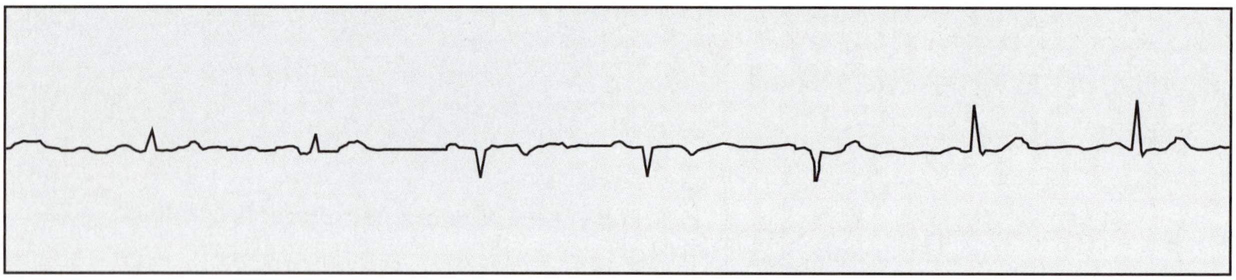

Compare both recordings above. What is different?

God has given doctors the ability to do heart surgery. This promise is for any of God's children who may have surgery.

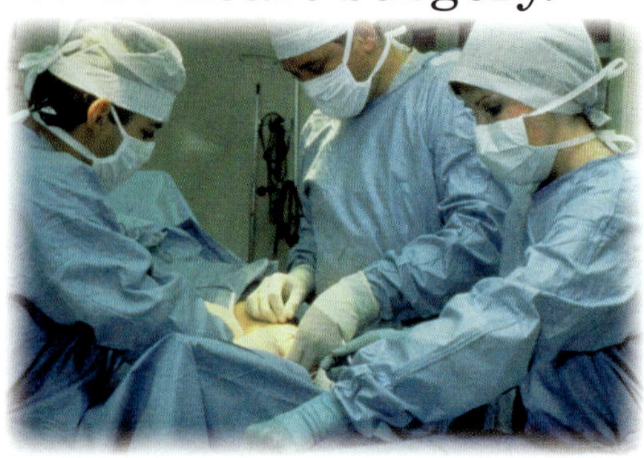

Be strong and courageous. Do not be terrified; do not be discouraged, for the Lord your God will be with you wherever you go. Joshua 1:9b

God has arranged the parts in the body, every one of them, just as he wanted them to be. I Corinthians 12:18b

1. Draw a line from each word to the part.

heart

blood vessel

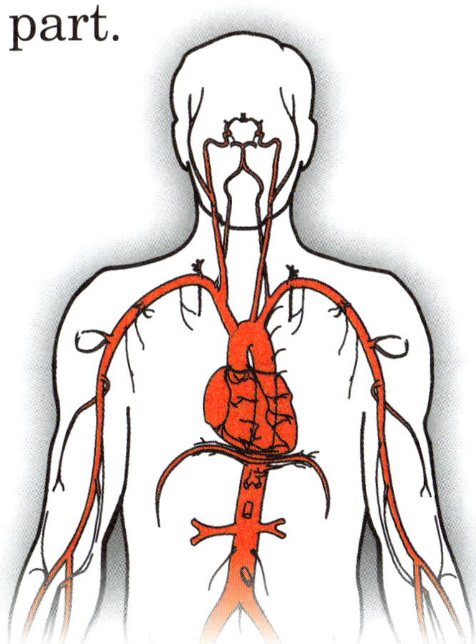

Use the words below to complete the sentences.

heart blood vessels

2. Blood flows to all parts of the body through

_______________ _______________.

3. Blood is pumped by an organ called the

______________________________.

4. Circle the pictures of foods that are good for your heart.

5. Fill in the circle to tell what should appear after a bandage is placed on a cut.

○ new skin ○ small scab ○ book

Chapter 9 Review

Read the creation web.

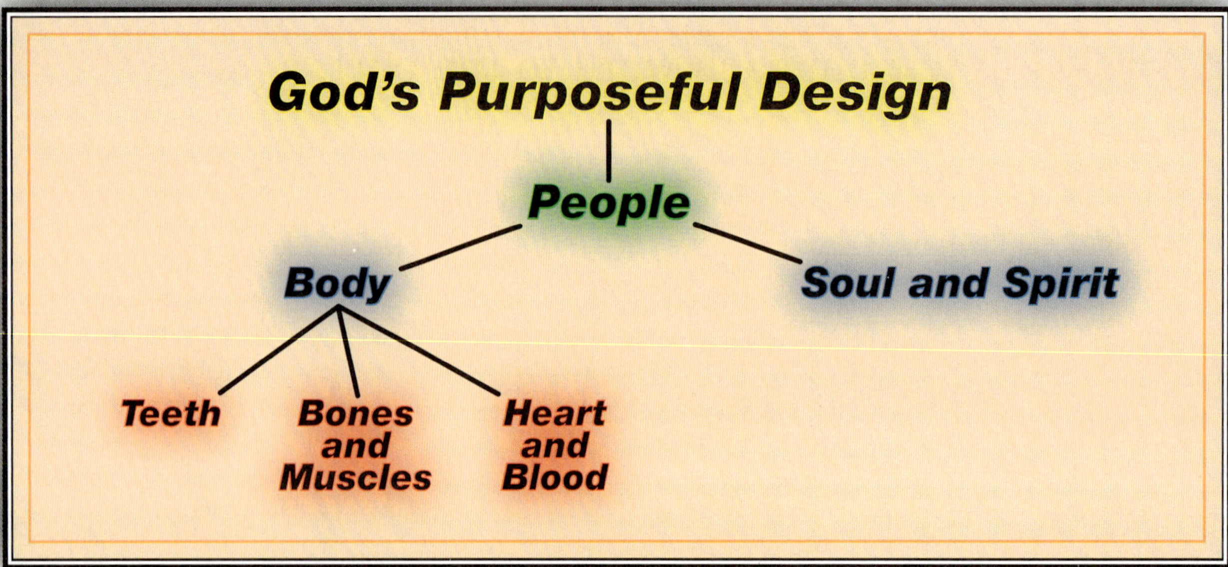

Journal about the Heart and Blood

Write one thing you learned about the heart.

I learned ______________________________

Air fills the bubbles.

My Booklet about Lungs and Air

Name ______________________________

Draw and color bubbles.

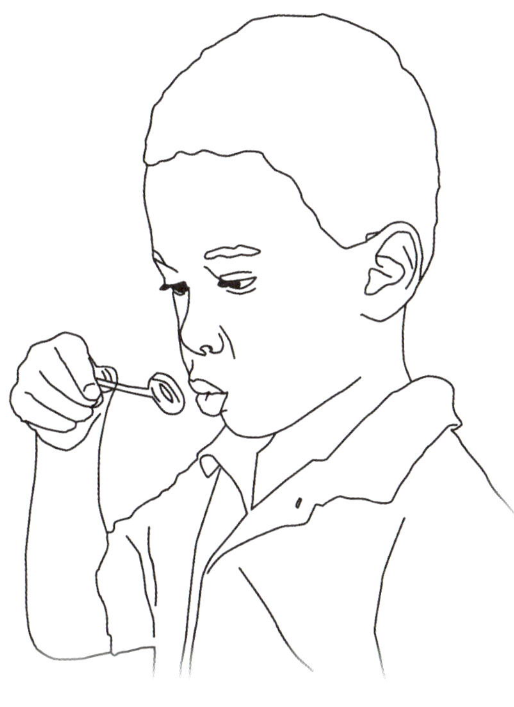

What Happens When You Breathe?

Lungs are organs that take in and let out the air you breathe.

Where is air found?

Air is all around you. It has no color, no shape, and no taste. Air is made of gases. Your body needs the oxygen in air.

Do you always breathe at the same rate?

Sit still. Then count the number of breaths you take.

Run in place. Then count the number of breaths you take.

Breaths in one minute	
sitting 	
running 	

Living things need air to stay alive. Circle ten living things.

What Happens When You Breathe?

You cannot see air, but you can see what it does.

Circle the object that is full of air.

Circle the flower that is falling apart by her breath.

Circle the soup that is being cooled by his breath.

Lesson Review

Circle the picture that answers the question.

1. Which part of your body helps you to breathe?

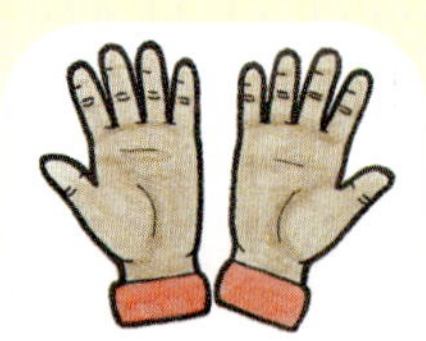

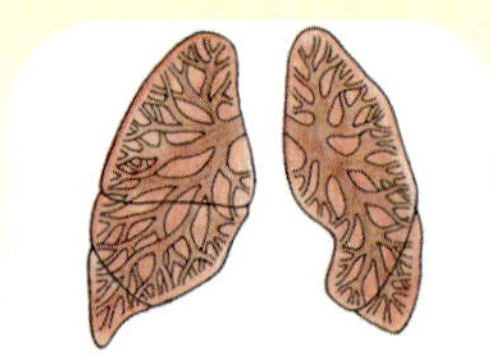

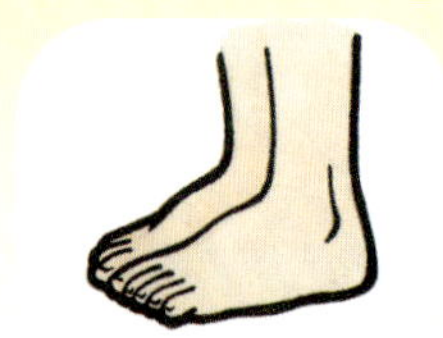

2. Which group of things needs air to stay alive?

Air enters the body through the nose or mouth.
It goes to the throat and down the windpipe.
It enters the lungs.

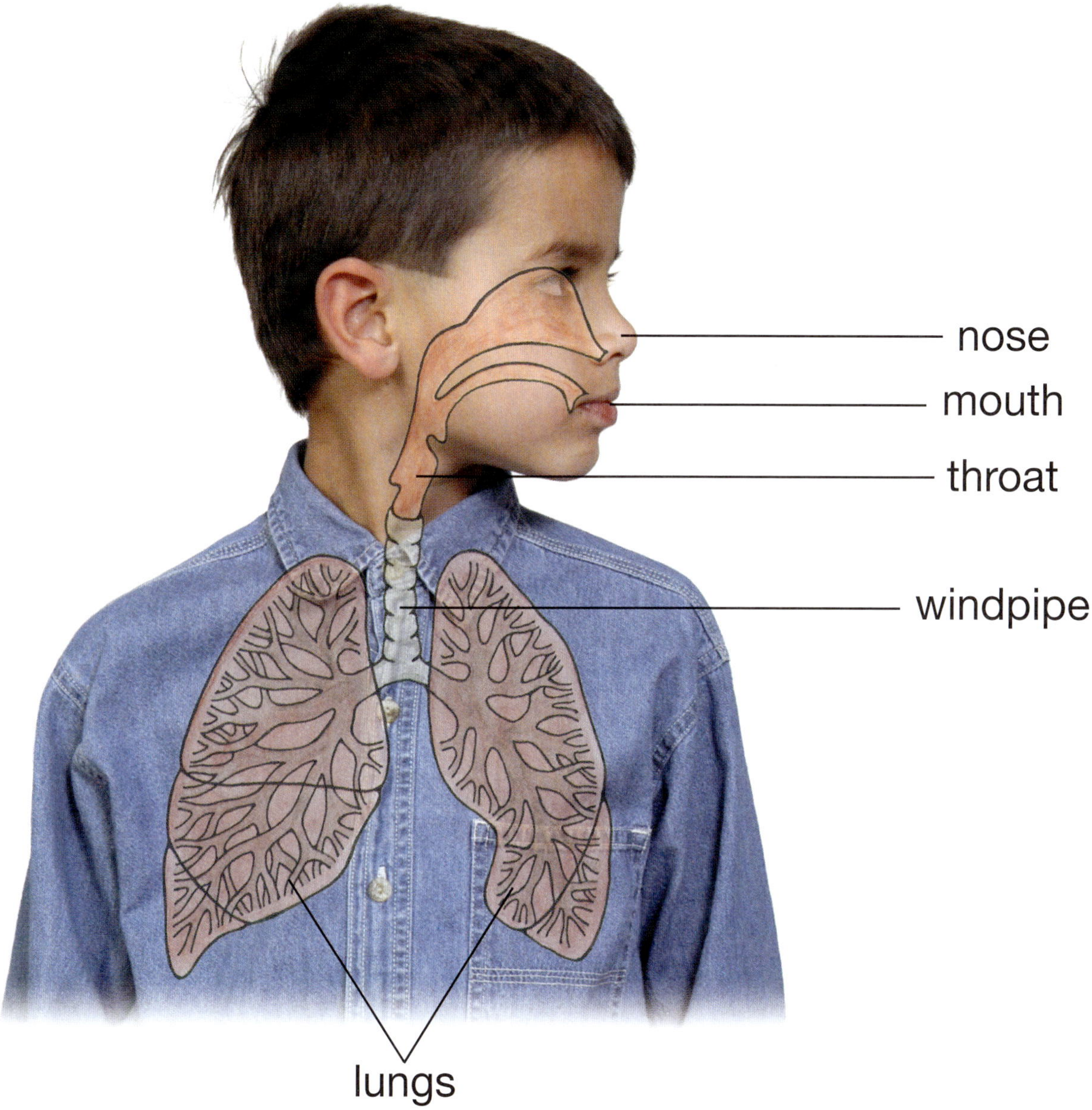

Inhale means to take air into the lungs.

Exhale means to let air out of the lungs.

What helps you breathe clean air?

Mucus catches the dust in the air. Tiny hairs move the mucus and dust out of the windpipe.

This helps your lungs to have clean air to breathe.

Draw tiny hairs.

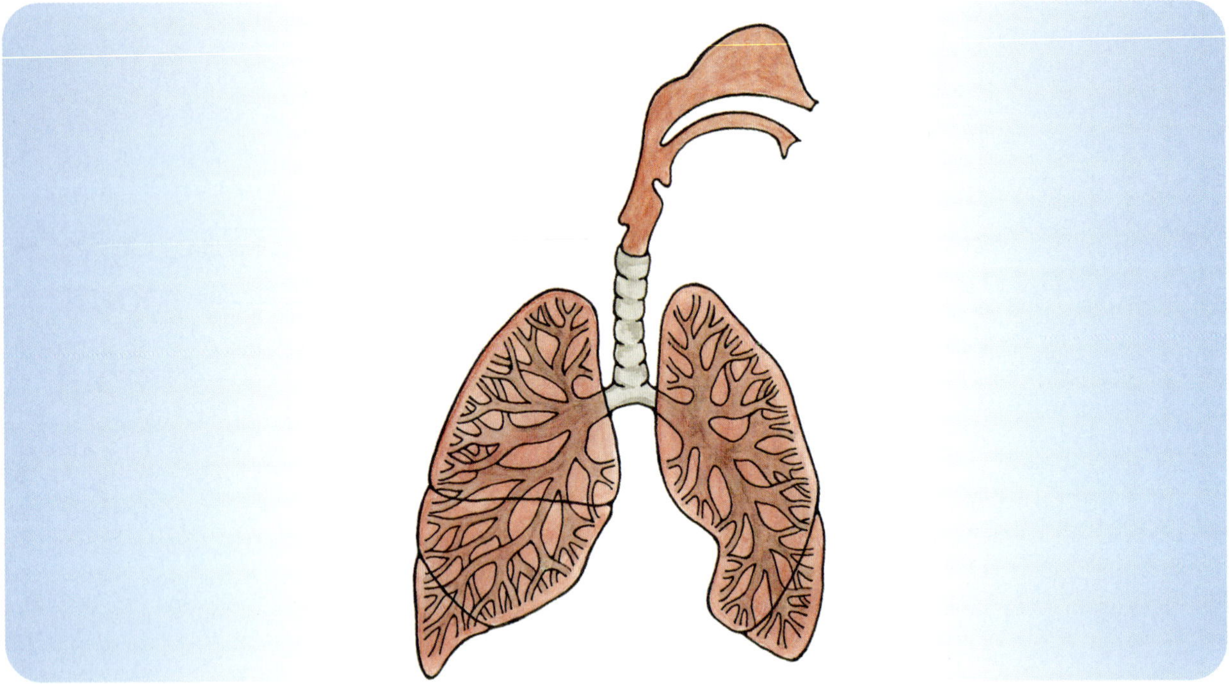

How much air do you breathe every minute?

Your body breathes about four quarts of air every minute. Here are four quart-sized bags.

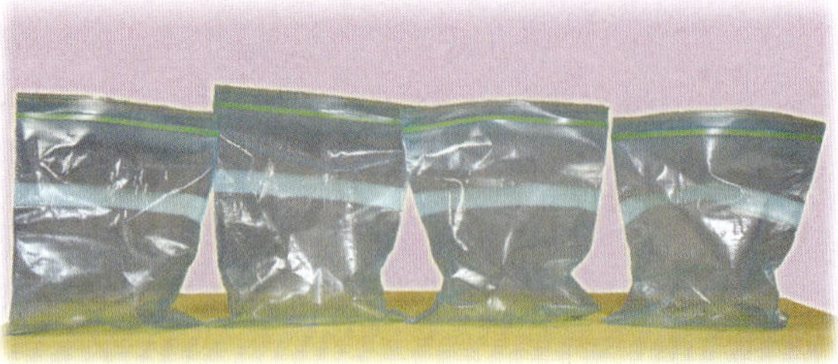

Lesson Review

Fill in the circle that answers the question.

1. What helps you to breathe?
   - ○ lungs
   - ○ legs
   - ○ arms

2. What helps to clean the air you breathe?
   - ○ soap
   - ○ tiny hairs and mucus
   - ○ water

You breathe to stay alive.
Oxygen helps keep your body working.

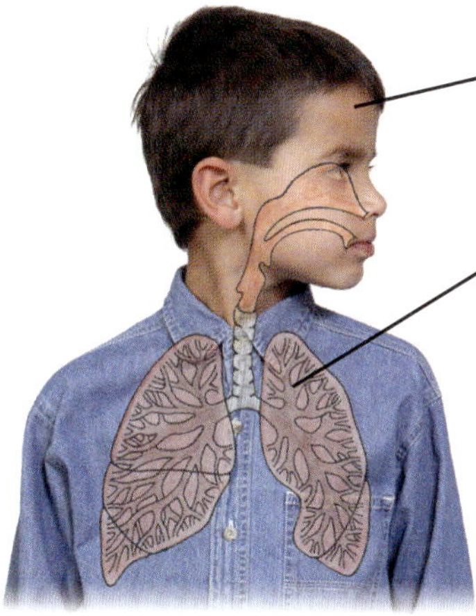

Your brain tells your body to breathe.

Your lungs take in air to get the oxygen your body needs.

Your lungs get rid of carbon dioxide that your body does not need.

In his hand is the life of every creature and the breath of all mankind.
Job 12:10

Do these animals breathe the same way?

lion

spadefish

Mammals use *lungs* to breathe. Fish use *gills* to breathe.

sparrow

beetle

Birds use *lungs* to breathe. Insects breathe through *tiny holes* in their skeleton.

Taking in Air

Reptiles use *lungs* to breathe.

chameleon

Some amphibians use *gills*, *skin*, and *lungs* to breathe.

frog

Earthworms breathe through their *skin*.

earthworm

Do living things breathe the same way or different ways?

Lesson Review

Fill in the circle beside the answer.

1. Two of these animals breathe with lungs. Which animal breathes a different way?

○  ○  ○ 

2. To stay alive, you breathe with your lungs. What tells your body to breathe?

○ brain ○ heart ○ bones

What helps you to talk or sing?

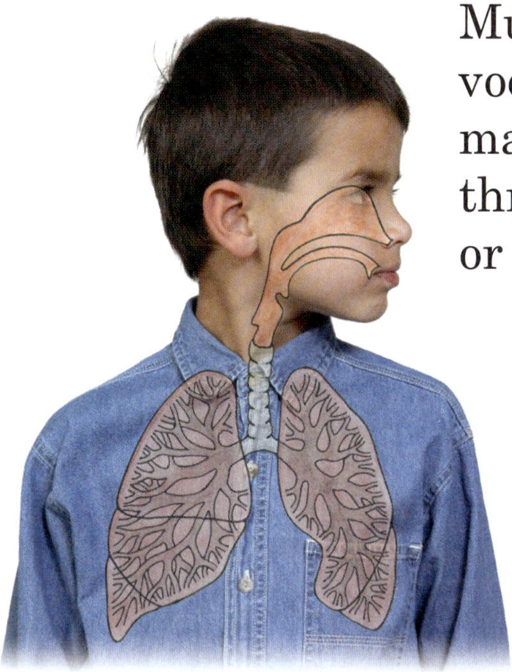

Muscles vibrate the vocal cords. The vocal cords vibrate your breath to make sound. Your mouth and upper throat form words to help you talk or sing.

What are some other things your breath can help you do?

Your breath can fill a balloon.

Your breath can blow bubbles.

More about Breathing

Your breath can blow out birthday candles.

Your breath is needed to play a flute.

Your breath can warm your hands.

Your breath can cool hot chocolate.

Does the temperature of her breath change?

Lesson Review

Your lungs help you breathe. Write about something your lungs can help you do as you breathe out.

What helps your lungs to be healthy?

The nose and windpipe were designed with hairs and mucus to help you breathe clean air.

Cover your mouth and nose with a tissue when you cough or sneeze. This keeps germs from spreading. Germs are so small they cannot be seen. But they are in the air you breathe.

A cold makes it hard to breathe. Wash your hands often if you have a cold or are around someone who has a cold.

Healthy Lungs

Breathing may be difficult if the tubes or airways to the lungs are not open enough.

Asthma makes breathing difficult.

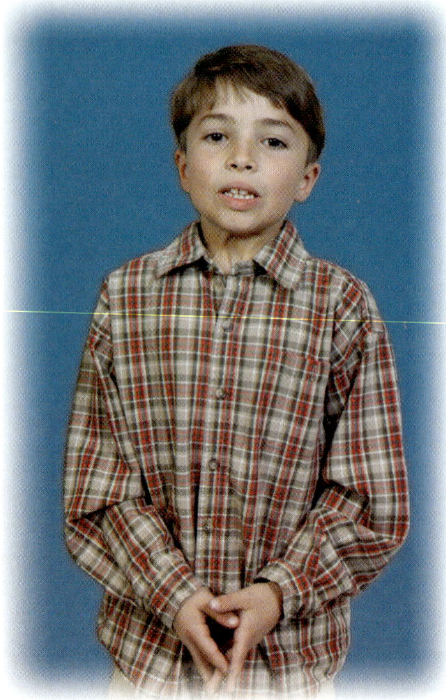

What are signs of asthma?

- A wheezing sound may be heard when one breathes out.
- It is hard to stop coughing.
- Breathing is difficult during and after play.

If you have trouble breathing, tell your parents.

Some people with asthma use an inhaler to help them breathe.

Lesson Review

Circle the word that answers each question.

1. What is in your nose that helps stop dirt from getting to your lungs?
   bones tiny hairs blood
2. What is one part of your body that is protected by washing your hands?
   feet teeth lungs

This is how healthy lungs look.

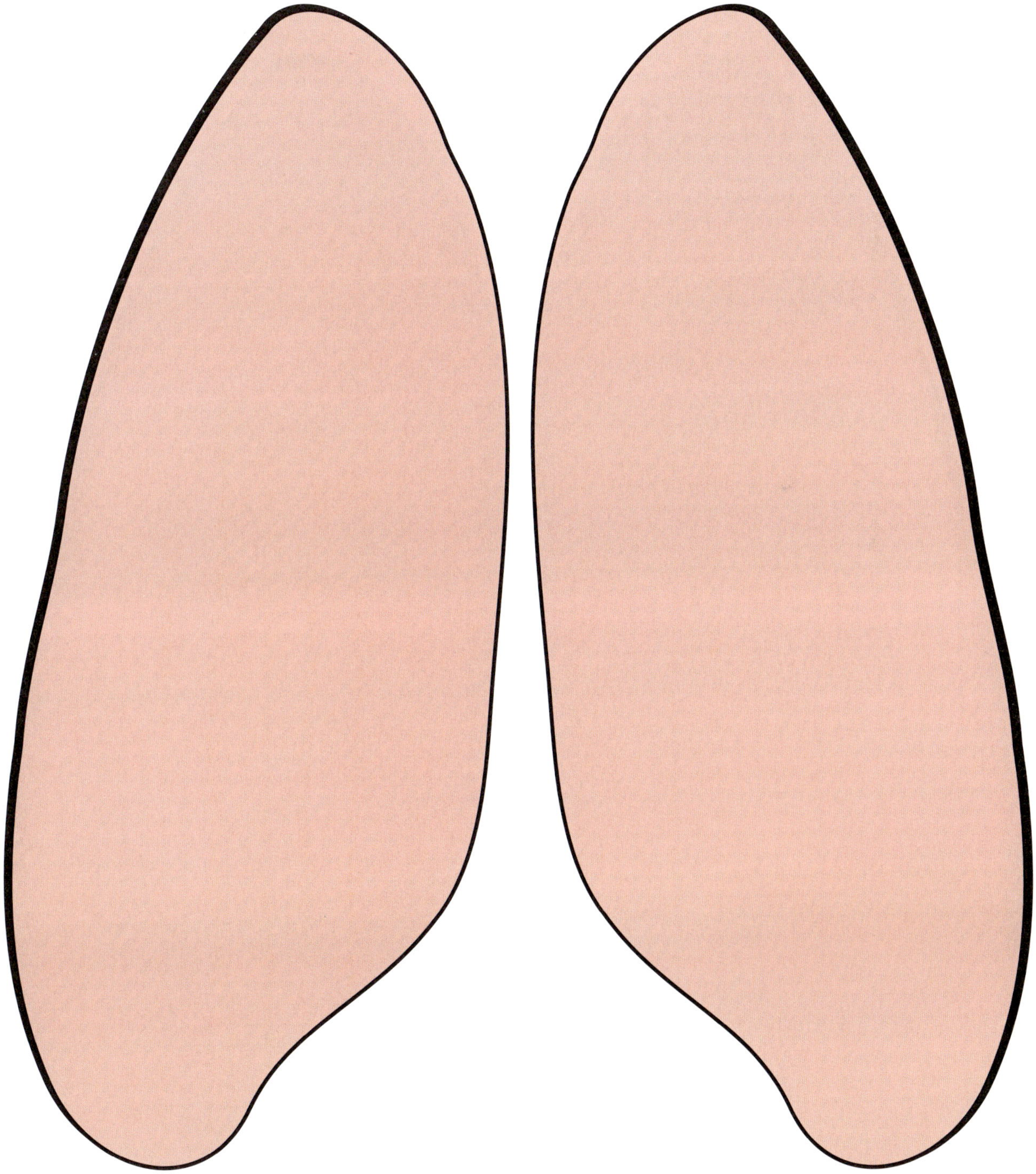

This is how unhealthy lungs look.

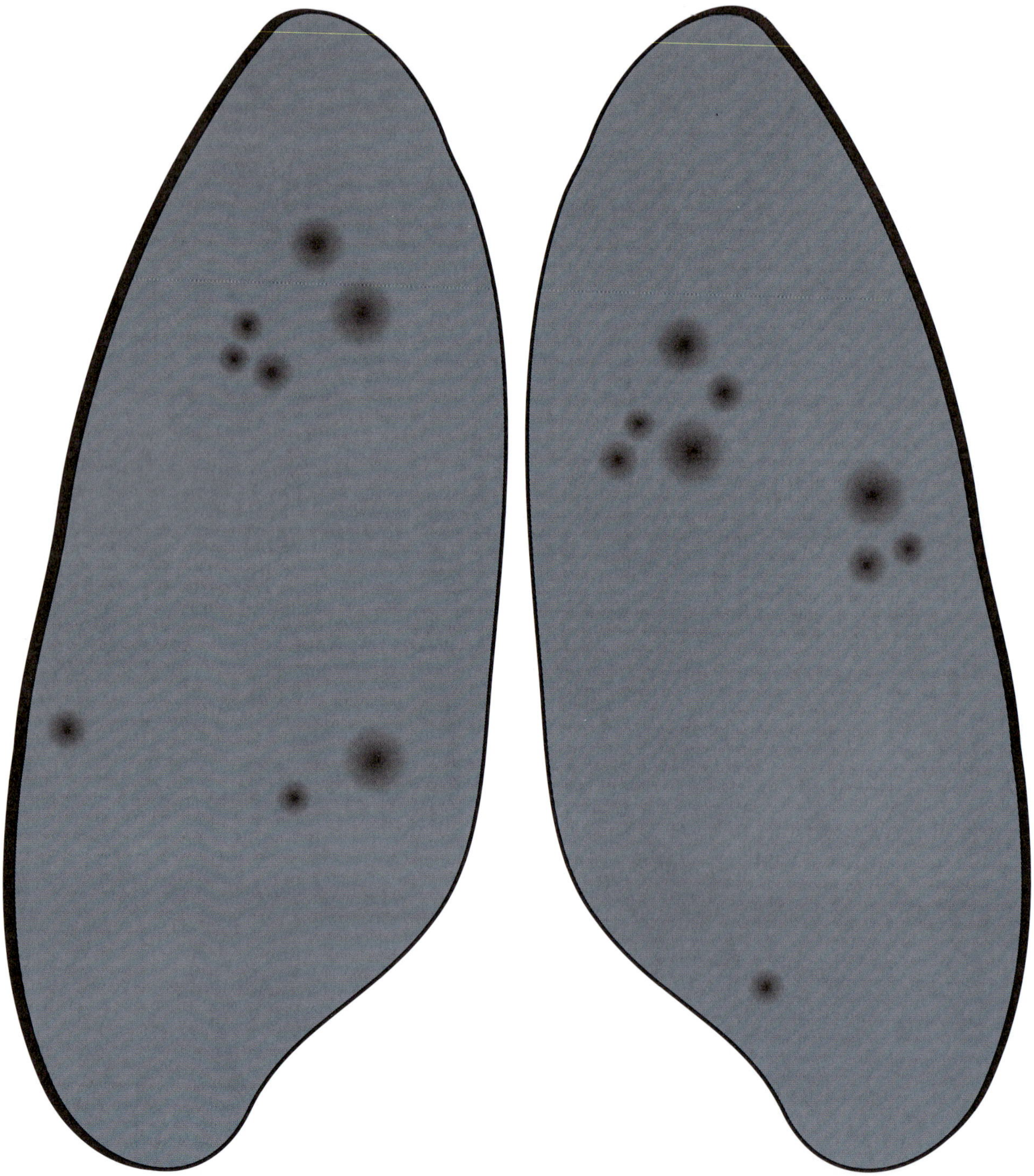

An astronaut uses an oxygen tank to be able to breathe in space.

A scuba diver uses an air tank to be able to breathe as the ocean is explored.

The firefighter uses an air tank to breathe while fighting a fire.

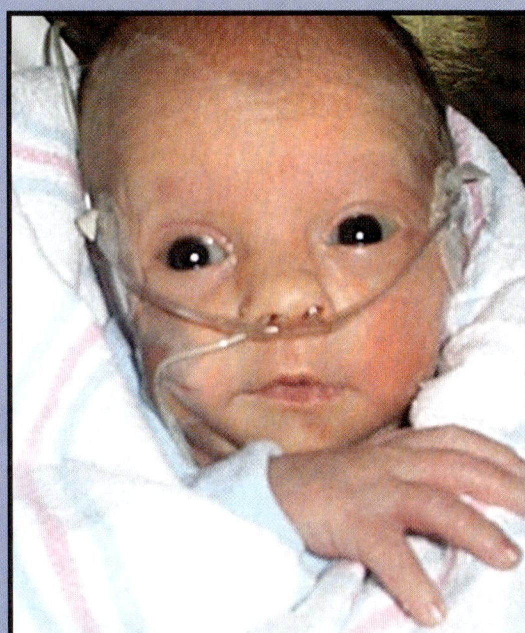

A baby that is born too early needs extra oxygen.

- [ ] Know the sound of a smoke detector.

- [ ] Do not touch matches or lighters.

- [ ] Have more than one escape plan. Go to the meeting place away from the house.

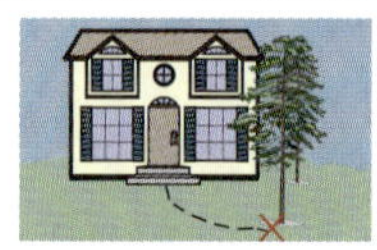

- [ ] Get out of the house if there is a fire! Once out of the house, do not go back inside.

- [ ] If the doors are blocked and you cannot get out, get on your hands and knees and stay below the fumes and smoke!

- [ ] Yell or wave something in front of a window.

- [ ] Stop, drop, and roll if your clothes are on fire.

- [ ] The firefighter is your friend! Firefighters wear a mask and an air tank to rescue people in a fire.

1. Living things need food, air, and water to stay alive. Is it food, air, or water that you cannot see? Circle.

food air water

2. Living things breathe different ways. Draw a line to match each animal with how it breathes.

lungs tiny holes gills

3. Use the words below to complete the sentences.

body brain move

Your ______________________________ tells your

______________________________ when to breathe.

The faster you ____________, the quicker you breathe.

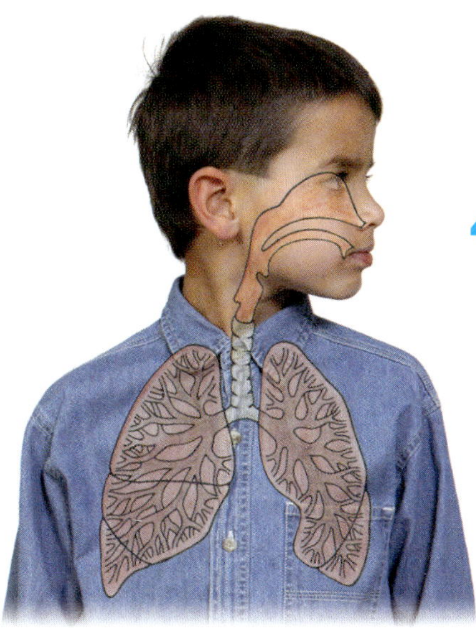

4. Use a yellow crayon to show air entering the nose, going down the throat and windpipe, and into the lungs.

Chapter 10 Review

Read the creation web.

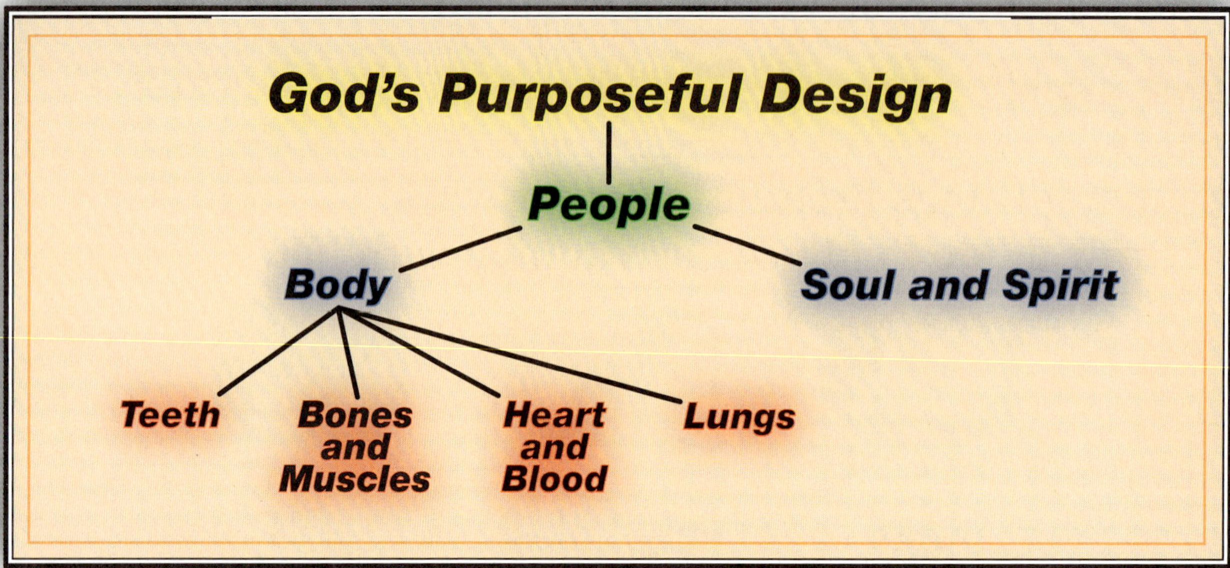

Journal about the Lungs and Air

Write what you know about fire safety.

I know

People need food.

My Booklet about the Stomach and Food

Name ______________________________

How many seeds are below? ______________ seeds

Kinds of Food

All living things need food. You need food. Some foods you eat come from plants.

The part you eat grows in the soil.

The part you eat grows out of the soil.

 The part you eat grows in a nut shell.

Some foods you eat come from animals.

and

come from fish.

 and 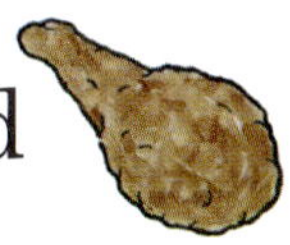 come from chickens.

 and 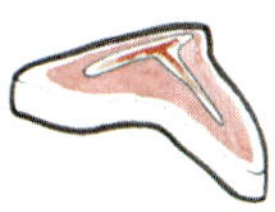 come from cows.

Milk comes from animals that have eaten plants.

Some foods are mixed and baked.

Lesson Review

Draw a line from the picture to the word that shows where each food comes from.

a plant an animal

What does this car need to be able to move? This car needs gasoline.

What does this toy car need to be able to move? This toy car needs a battery that works.

What do people need to be able to move? People need food. Food gives your body energy to work and grow.

Food Is Fuel

Draw a picture of what you ate for breakfast.

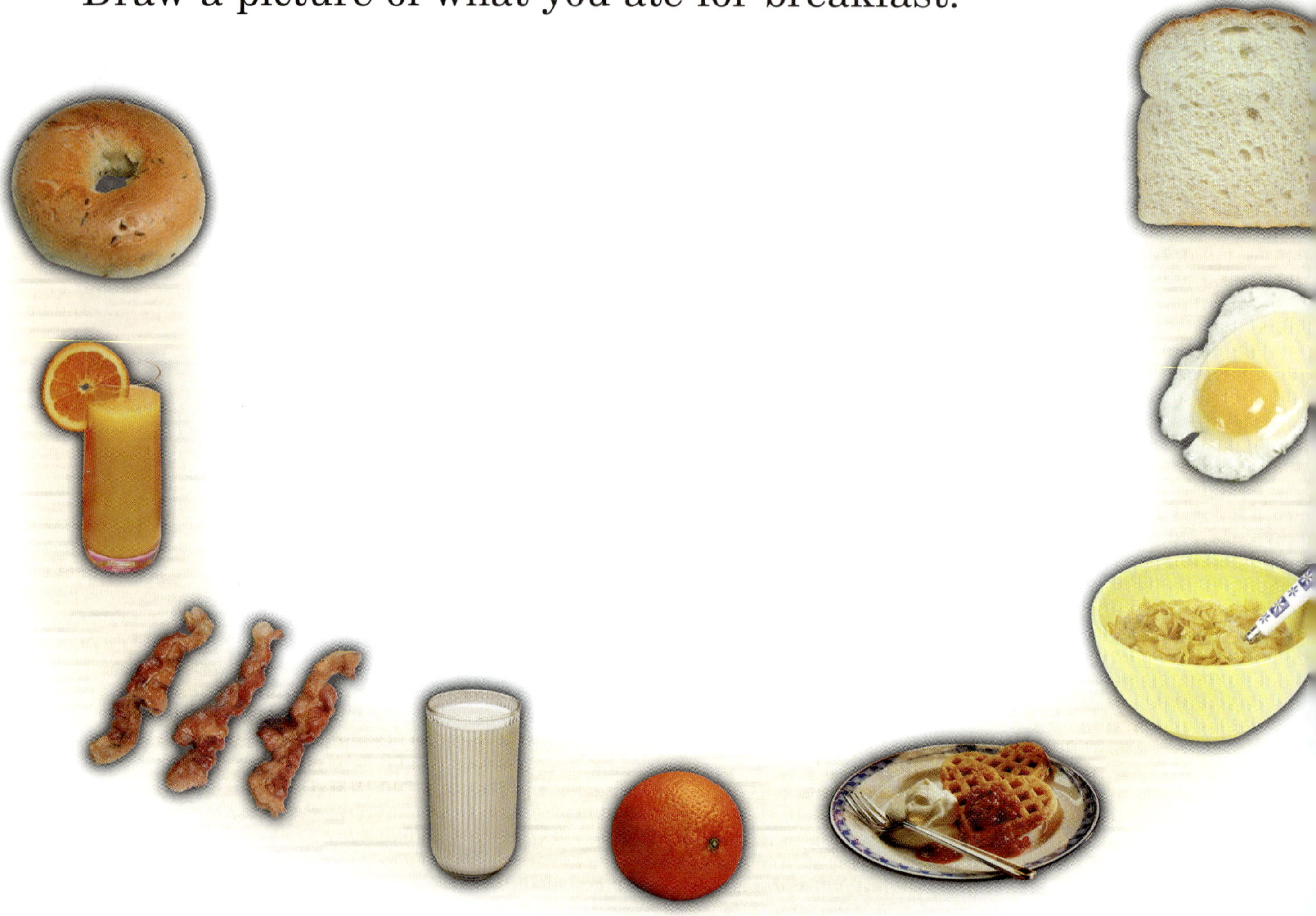

The foods you ate for breakfast gave you energy to work and grow. Make a list of work you have done today.

1. walked to class
2.
3.
4.

What would you like to eat? Use this guide to know which foods help you grow strong and stay healthy.

Many years ago people built pyramids. They look tall and strong.

Use the Food Guide Pyramid to grow strong.

Food Guide Pyramid

Draw a food for each group. Write the number of servings per day you should eat.

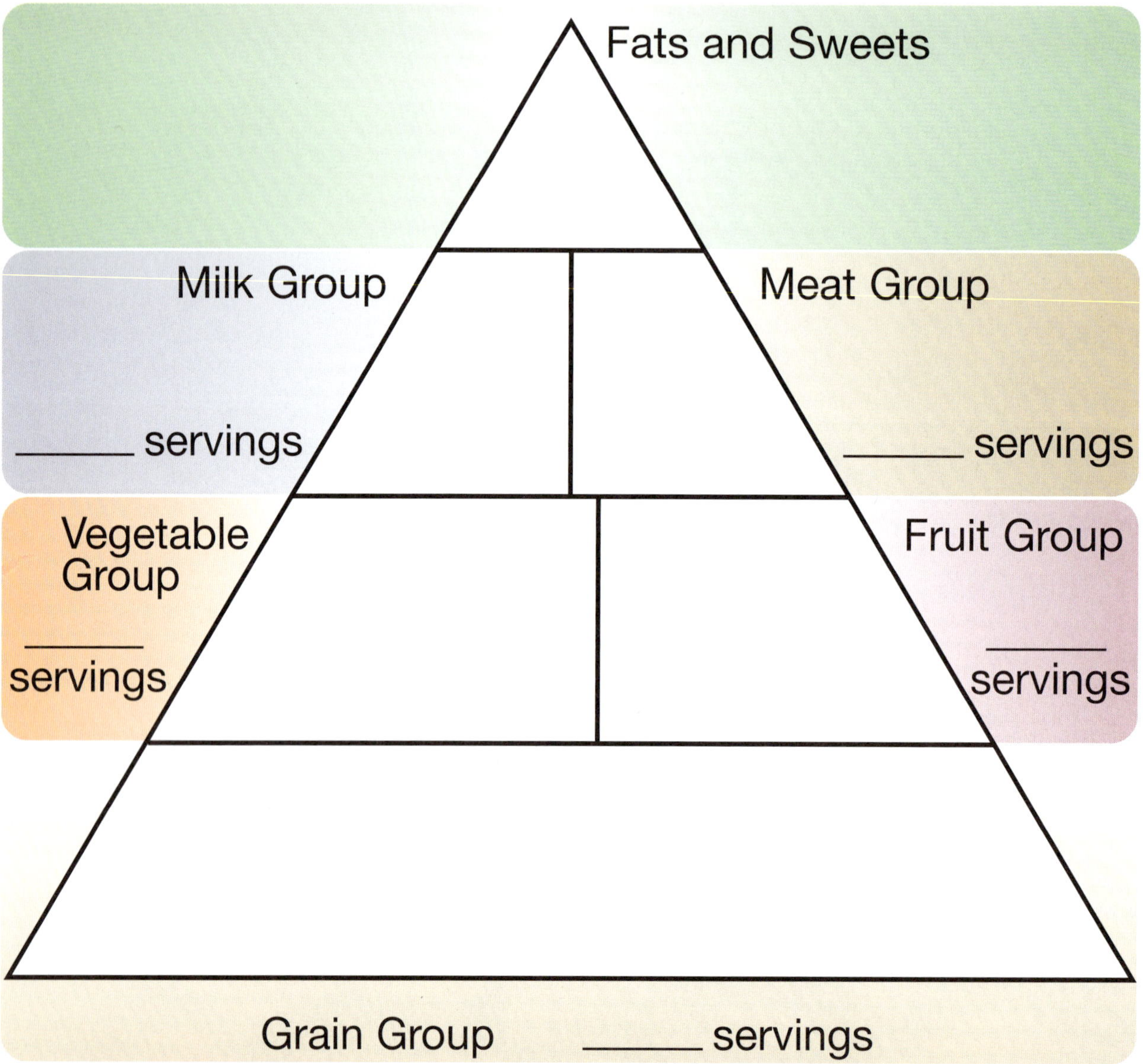

Lesson Review

Circle the more healthy meal.

God designed your body to break down food into tiny pieces.

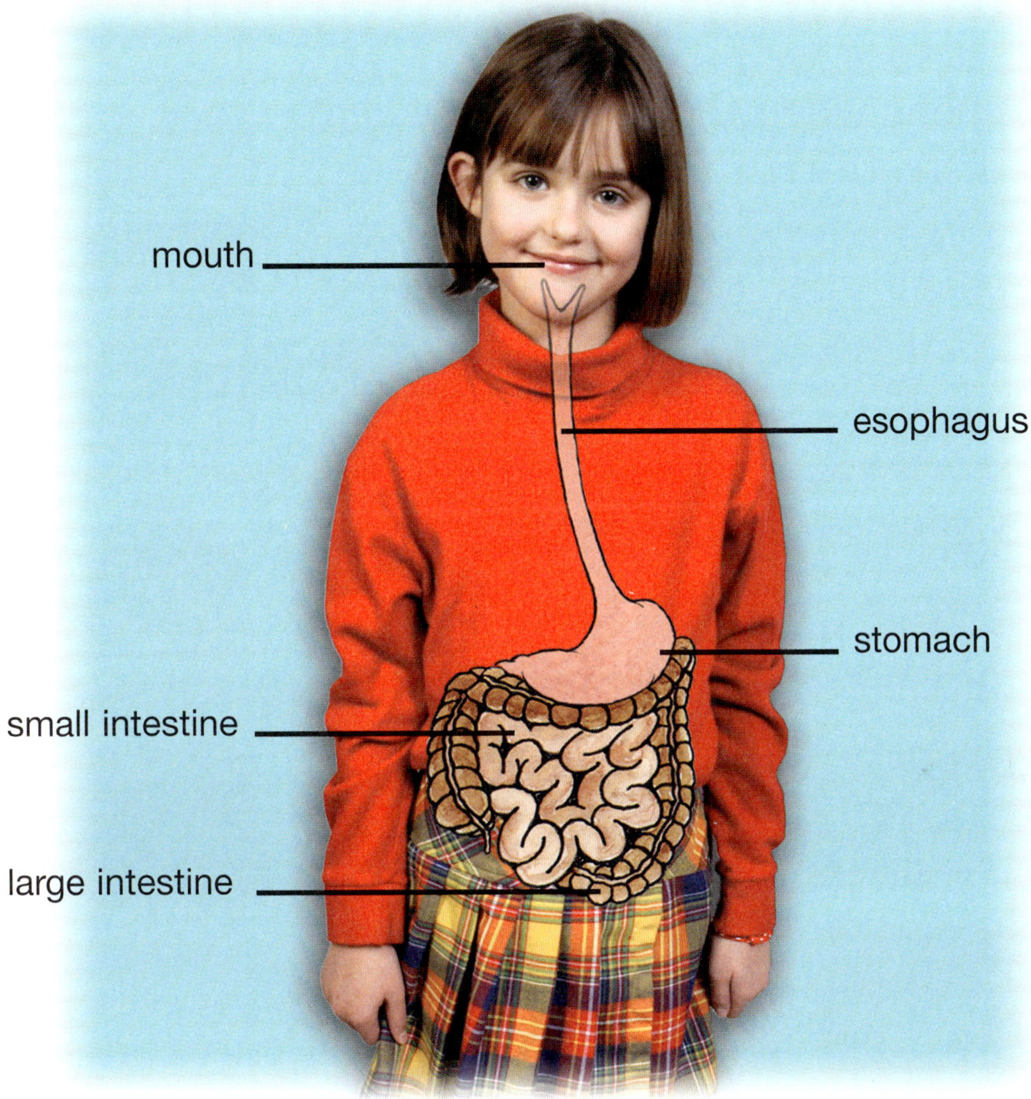

You chew food with your mouth. The food you swallow goes down the esophagus to your stomach. Your **stomach** is an organ that breaks down food.

Then the food passes into the small intestine. Nutrients from the food go to all parts of your body.

The foods that cannot be broken down are wastes. They continue on into the large intestine.

Where Does Food Go?

How does your mouth work?

Observe the sugar cube. 

What happens when the cube is crushed?
Could your teeth crush it into pieces?

What happens when the cube is placed in water?
Could the saliva in your mouth dissolve it?

How does your esophagus work?

Pour water in a tube.

What happens?

How does your stomach work?

Your stomach mixes food the same way your hands squeeze the bread.

The intestines are now ready to use the tiny pieces of food to give you energy.

Lesson Review

Number the parts in the order that food follows.

stomach___ mouth___ intestines___ esophagus___

Mark an *X* on each picture from the food group.

1. fruit   

2. milk 

3. vegetable  

4. grain 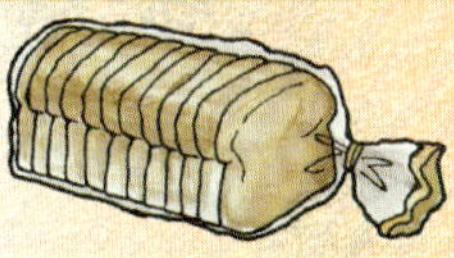  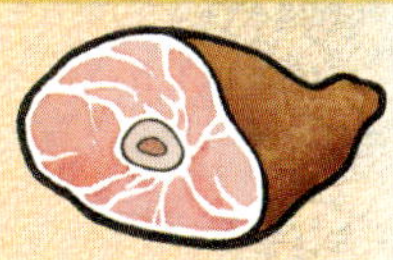

5. meat  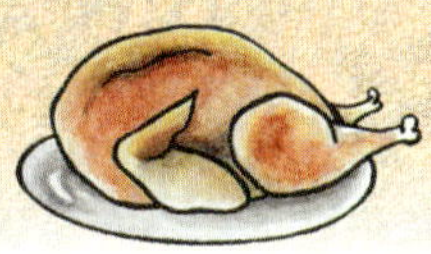 

Outline the parts of the body with these colors.

6. 
esophagus

7. 
stomach

8. 
large intestine

9. 
small intestine

So whether you eat or drink or whatever you do, do it all for the glory of God.
I Corinthians 10:31

Chapter 11 Review

Read the creation web.

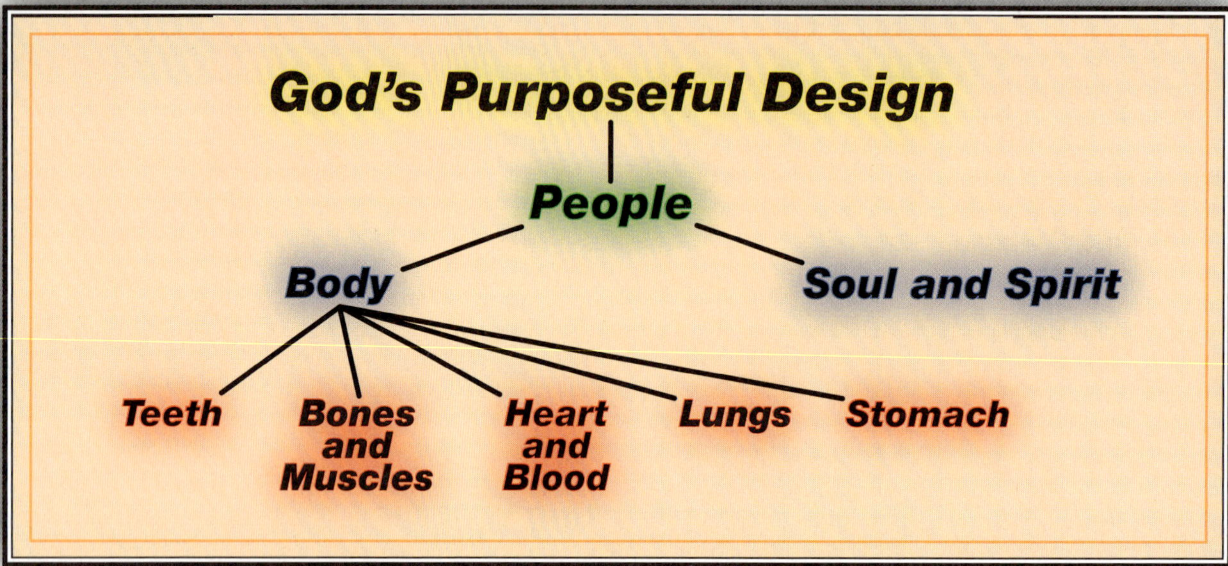

Journal about the Stomach and Food

Write about a food you like. Name its food group.

Fall is one of the four seasons.

My Booklet about Seasons

Name ______________________________

Color.

A season is a time of year.

What are the different seasons?

Winter is the coldest season. It can be rainy and snowy. People wear heavy coats outside.

Many deciduous trees lose their leaves once a year. They have no leaves in winter.

Some animals like bears hibernate. To hibernate means to spend the winter in a deep sleep.

Spring is when temperatures start to warm up outside.

Trees begin to grow new leaves and flowers.

Some animals like the robin return to their summer home.

Farmers plant crops.

Summer is the warmest season.

Plants and flowers grow.

Animals are active.

People enjoy being outdoors.

What Are Seasons?

Fall is when temperatures start to get cooler outside.

Deciduous trees lose their leaves.

Migrating animals like the arctic tern go to warmer places for food.

Some animals like beavers collect and store food for winter.

Are the seasons where you live like the ones in this lesson? Discuss the seasons where you live.

Why are there seasons?

The sun warms Earth. Earth travels around the sun. Earth is tilted as it travels.

The part that is tilted toward the sun is having summer.

The part that is tilted away from the sun is having winter.

Lesson Review

1. What is the weather like in each season?
2. How is fall different from spring for migrating animals?
3. Why do plants grow well in the summer?

How do the seasons help an orange to grow?

In spring, the orange tree has green leaves and white blossoms that smell sweet.

In summer, the petals of the flowers fall off and small green fruit is left to grow.

In fall and winter, oranges are picked and eaten.

"As long as the earth endures, seedtime and harvest, cold and heat, summer and winter...will never cease."
Genesis 8:22

Which do you enjoy the most—a glass of orange juice, an orange popsicle, or orange slices?

Take a class vote. Color one space on the bar graph for each vote. A bar graph is used to collect data.

 **juice**										
 **popsicle**										
 **slices**										

Seasons and Plants

The oak tree is deciduous. It loses its leaves in the fall. It grows new leaves in the spring.

The white pine tree is an evergreen. The needles stay on this tree all year long.

How is the oak tree similar to the pine tree?

How is the oak tree different from the pine tree?

Spring, summer, fall, and winter happen every year because of the change or cycle of seasons. Events that repeat in a certain order over and over again are called a cycle.

Lesson Review

1. Why does a plant start to grow in the spring?
2. Tell about something that an adult might plant in your yard in the spring.

When something begins and ends at the same place, it has traveled in a cycle.

The hands on a clock begin and end at number twelve.

The earth travels around the sun from one season to the next every year.

What does the cycle of seasons mean to the Shetland pony?

In fall the pony starts to grow a thicker coat of hair.

In winter it has a very thick coat of hair to keep warm.

In summer it has thinner hair to keep cool.

In spring it sheds its thick coat of hair.

How do these animals find shelter in the winter?

Insects may hide under the bark of trees.

Snakes may hide in underground holes.

Draw the place where a pet finds shelter in the winter.

How do animals stay alive in the winter?

The elk moves from place to place to find food that is not covered by snow.

The chipmunk eats extra food in the fall that is stored as body fat and sleeps all winter.

The snowshoe rabbit grows new fur in the fall to keep warm and is camouflaged in the winter.

Decide how each animal is designed to stay alive in the winter. Fill in the correct space in the chart.

elk			
chipmunk			
snowshoe rabbit			
	migrates	stays active	hibernates

Read the chart to complete each sentence.

1. The elk ____________________________.

2. The chipmunk ____________________________.

3. The snowshoe rabbit ____________________________.

What do the seasons mean to farmers?

Some farmers plant winter wheat in the fall. It grows over the winter. That is why it is called winter wheat. It is harvested in spring.

Wheat helps people to have bread.

"I will send you rain in its season, and the ground will yield its crops." Leviticus 26:4a

What do the seasons mean to people on vacation?

Some people enjoy the winter sport of skiing.

Others like the warmth of summer at the beach.

What do the seasons mean to children in school?

Recess may be inside on a cold winter day.

Spring break means a few days out of school.

What do the seasons mean to Christians?

The birth of Christ is celebrated at Christmas which is in the winter.

Christ arose at Easter which is in the spring.

Seasons and People

Tally means to make a count of.

| is a tally mark that stands for 1. Count by ones.

卌 stands for 5. Count by fives.

This cycle of seasons shows each person's favorite season.

Draw the tallies and tell how many.

Season	Tally	How many?
winter	||||	4
spring	卌 |||	8
summer	______	______
fall	______	______

Which season is the favorite? ______

Which season is the least favorite? ______

Everything around you is either a solid, a liquid, or a gas.

A **solid** has a definite shape. A sheet of paper and a hand lens are solids. They keep their shape after being used.

A **liquid** takes the shape of its container. You can pour a liquid. Pour a tall glass of fruit punch into a bowl.

How does the shape of the fruit punch change?

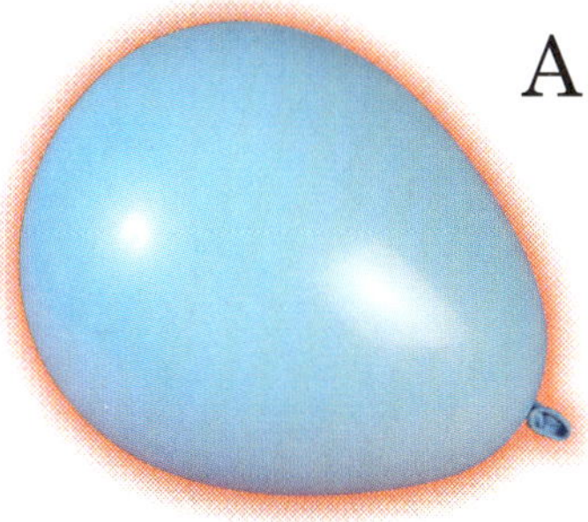

A **gas** fills what it is inside of.

Water can be a solid, liquid, and a gas.

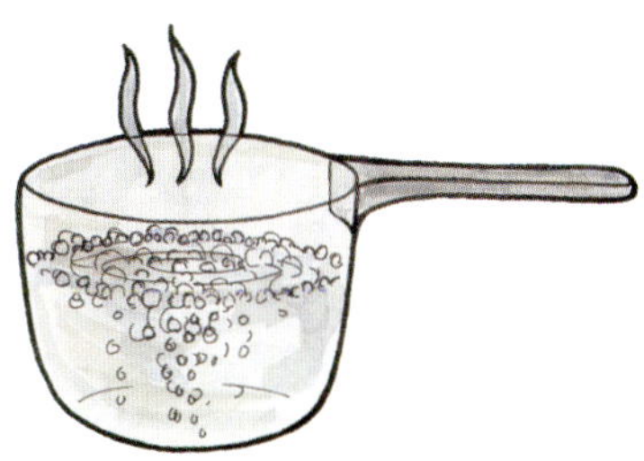

The water in a pond can look different each season.

In fall the pond starts to freeze again when the temperature gets cold.

In winter the pond may be frozen solid if it is cold enough.

In summer the pond is warm enough to swim in.

In spring the edge of the pond starts to thaw.

Temperature is how hot or cold something is. Read the number where the red line stops on a thermometer. That is the temperature.

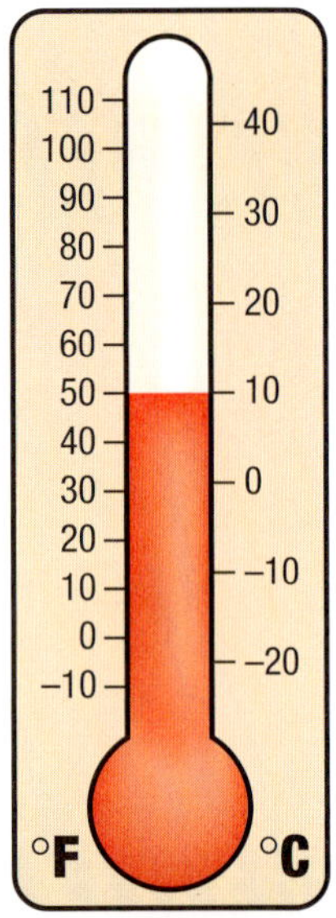

Find the temperatures inside. Then place the thermometer, the water, and the soil outside.

inside	outside (2 hours later)
classroom ________	outside ________
cup of water ________	cup of water ________
cup of soil ________	cup of soil ________

Experiment

Materials

salt sand
water cups

Skills

predict
observe
write a conclusion

Problem What happens to a cup of salty water and a cup of sandy water when they are placed in the open air?

Predict I think the water will

Plan How can you test this prediction?

Where Does the Water Go?

Experiment Pour salt into a cup of water. Stir.

Pour sand into a cup of water. Stir.

Mark the water level on each cup. Wait a day.

Observe Is the water level on each cup the same as it was at the beginning of the experiment?

Collect the data The water level is not the same as it was at the beginning. Some water evaporated. It changed to a gas. It went into the air. You cannot see the gas called water vapor.

Where is the salt?

Where is the sand?

Conclusion Was your prediction right?

Have you seen these kinds of clouds?

Cumulus clouds are puffy on the top. It will be a nice day.

Stratus clouds are layers of clouds. It might rain.

How does rain happen? Read numbers 1 through 4.

The Water Cycle

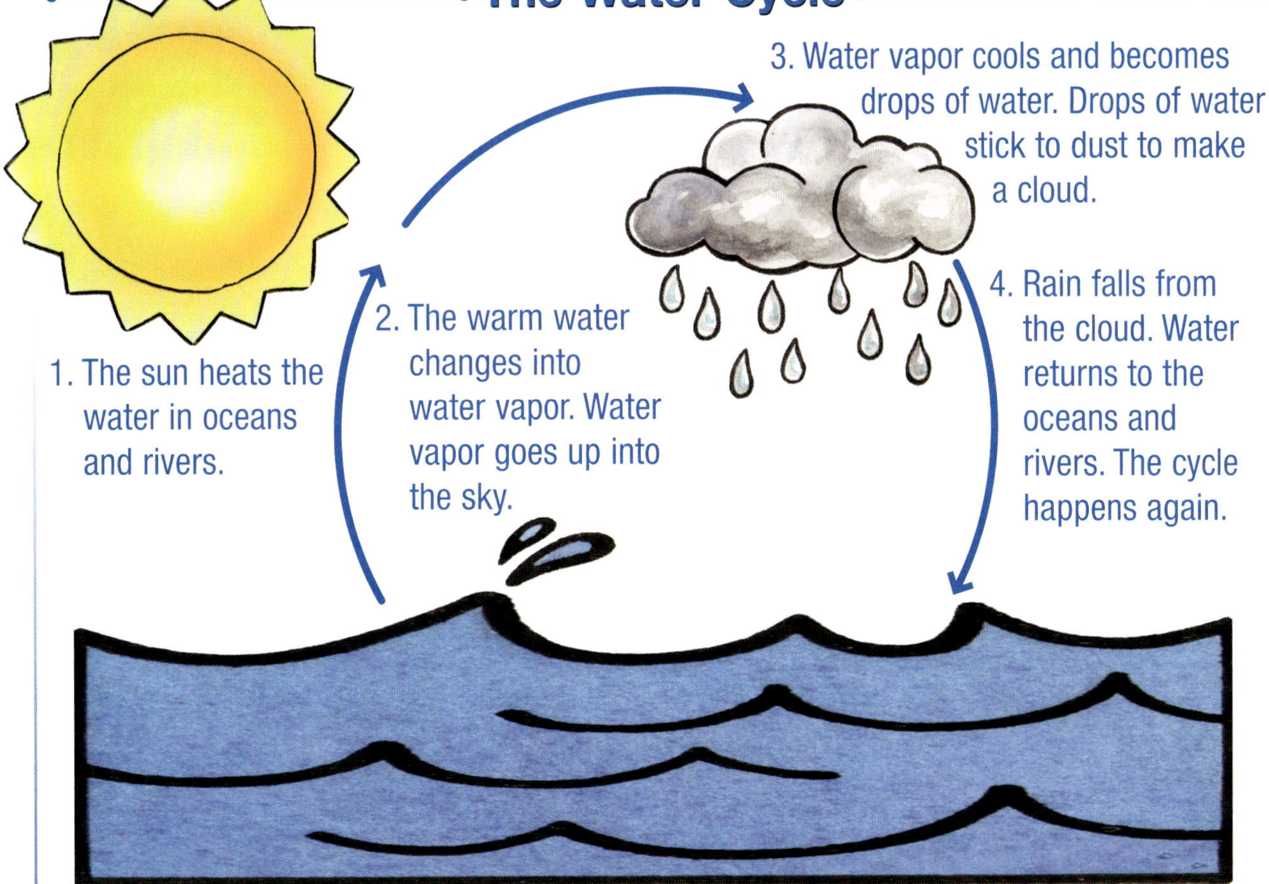

How does a raindrop become snow, hail, or sleet?

When water vapor freezes, it falls as snow.

A raindrop in a storm cloud can freeze. It falls as hail in the summer.

When rain freezes in the winter, it can fall as sleet.

He says to the snow, 'Fall on the earth,' and to the rain shower, 'Be a mighty downpour.' Job 37:6

How can you stay safe in a storm?

A thunderstorm has wind, rain, lightning, and thunder. Keep away from trees.

National Oceanic and Atmospheric Administration/Department of Commerce

A tornado has spinning winds and a funnel-shaped cloud. Keep away from windows.

National Oceanic and Atmospheric Administration/Department of Commerce

A hurricane has powerful winds. It can have rain, thunder, and lightning. Find a safe place in your home or nearby shelter.

National Oceanic and Atmospheric Administration/Department of Commerce

Lesson Review

1. What does water vapor become?
2. How can people stay safe in a thunderstorm? tornado? hurricane?

Tech Connect

A meteorologist studies weather conditions to make forecasts. Doppler radar may be used.

Doppler Radar Tower

Panels are replaced. The dome protects the dish.

The rotating radar dish is inside the dome.

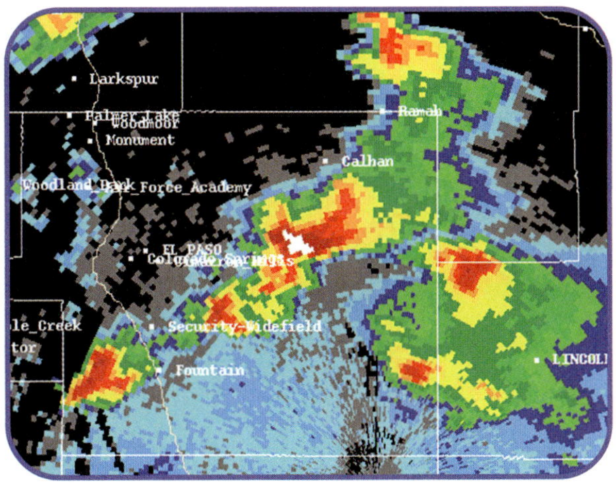

The yellow and red colors show that it is raining.

Tech Connect

What are some tools that measure the weather?

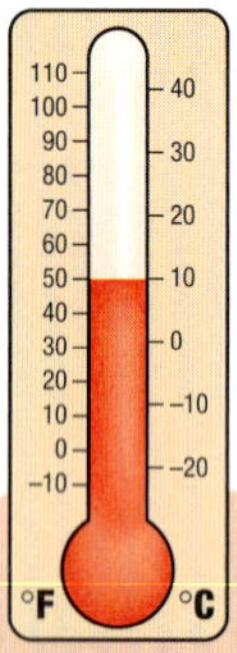

A thermometer measures temperature.

Measure and record the temperature three different times during the day.

	First	Second	Third
Temperature	______	______	______

The temperature ranges from __________ to __________.

What is the difference between the high and the

low temperatures? ____________________________

A rain gauge measures rainfall.

Make your own rain gauge. Set a cup outside in the rain. After it rains, measure the amount of rain with a ruler.

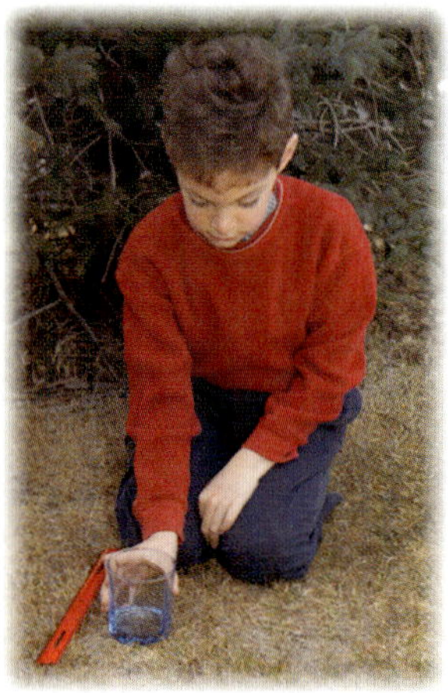

A windsock gives the direction of the wind blowing through it. Which direction is the wind blowing?

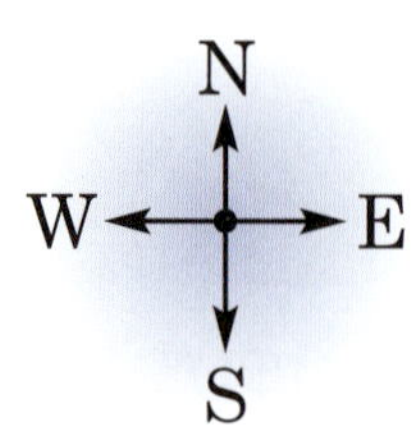

Circle the word that completes each sentence.

1. A season is a time of __________.

   class year lunch

2. Earth tilts away from the sun during __________.

   winter summer thunderstorms

3. A __________ takes the shape of its container.

   solid cycle liquid

4. When warm water changes to a gas, it __________.

   evaporates sinks melts

Use the bar graph. A cup of ice cubes melted. Who predicted the greatest number of ounces of water?

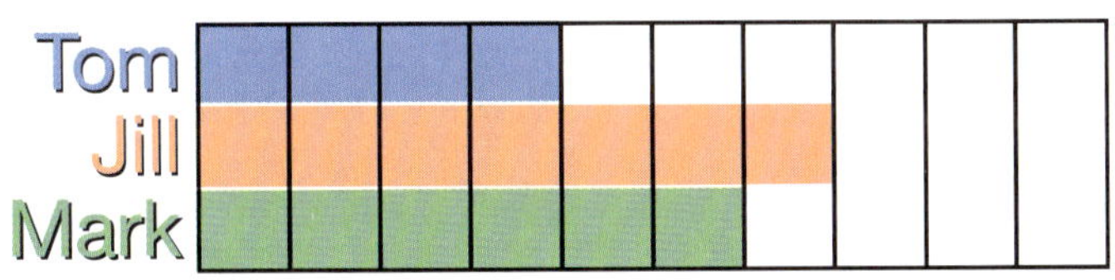

____

Complete each tree to show how it looks each season. You may use crayons.

Read the creation web.

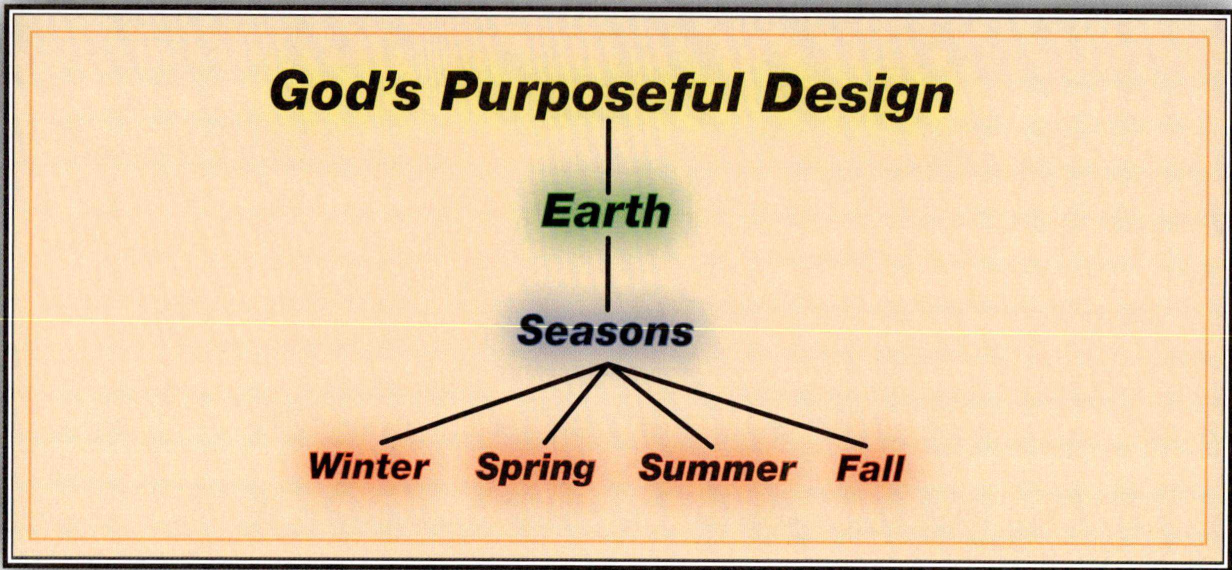

See! The winter is past; the rains are over and gone. Flowers appear on the earth. Song of Songs 2:11–12a

Journal about Seasons

Make up a story about the clouds.

From earth to space and back again.

My Booklet about Space

Name ____________________

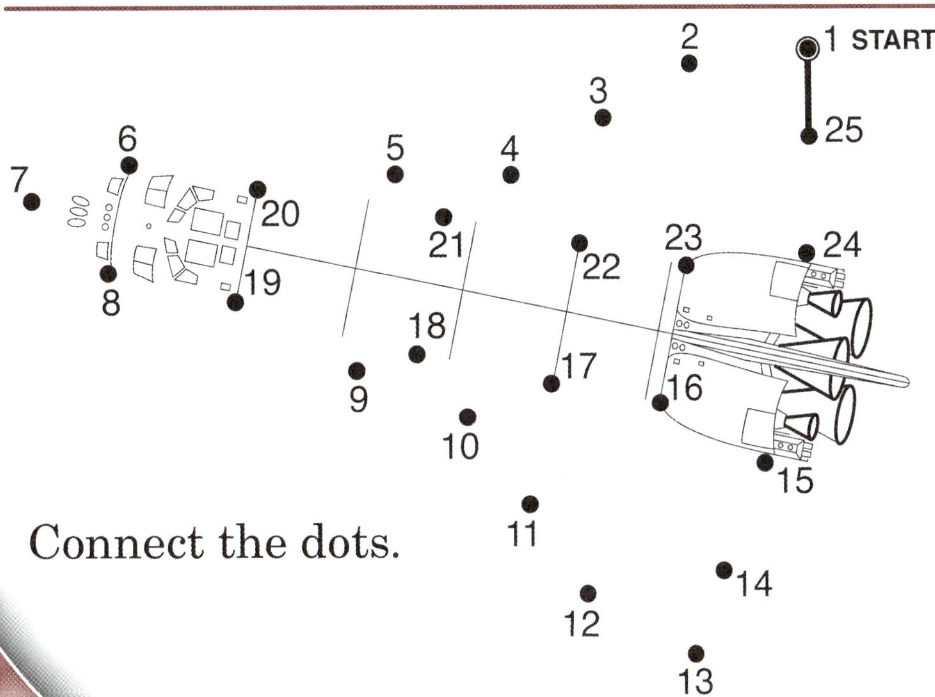

Connect the dots.

What causes day and night?

Earth spins, or rotates, once a day in a complete cycle.

Pretend a round balloon is Earth. Draw a line around the center for the equator. Tape a paper person on one side. Spin the balloon in one complete circle.

Pretend a flashlight is the sun. Shine the flashlight on the center of the balloon.

The side of Earth facing the sun is having day. The side of Earth facing away from the sun is having night.

Keep Earth spinning. This is the cycle of day and night.

Which is larger, the sun or Earth?

The sun is much larger than Earth. But the sun is so far away, it looks smaller.

Look at the airplanes. They are far away. They look smaller.

Look at the airplanes. They are near. They look larger.

As Earth spins, the sky changes.

The sun is in the east each morning. Is the sun *high* or *low* in the sky? Circle.

The sun looks like this at noon. Is the sun *high* or *low* in the sky? Circle.

The sun is in the west each evening. Is the sun *high* or *low* in the sky? Circle.

Why are there shadows?

Light comes from the sun. Light cannot go through solid things like a house. The house blocks the light from hitting the ground and makes a shadow. Can you see the shape of the shadow is like the shape of the house?

This shadow is short. The sun is high in the sky. It is noon.

This shadow is long. The sun is low in the sky. It is morning or evening.

Lesson Review

1. What causes day and night?
2. Circle the shadow.

Earth has air around it that you cannot see. This air is called the atmosphere.

How does the atmosphere protect Earth?
It blocks some harmful sun rays.
It keeps Earth from being too cold.
It stops most meteors from reaching Earth.

Space is outside Earth's atmosphere.
What could be there?

Where Is Space?

How do people observe space?

People look up into the sky to see the sun, moon, and stars. They appear small.

A telescope can make things that are far away appear closer and larger.

What do you think this girl sees when she looks through the telescope? Write about it.

Lesson Review

Fill in the circle that answers each question.

1. What is around Earth that protects it?
   ○ space ○ atmosphere ○ meteorites
2. Which tool makes far away things appear closer?
   ○ ruler ○ thermometer ○ telescope
3. Which could be seen in space?
   ○ stars ○ houses ○ living things

The sun is a star. It is the closest star to Earth. That is why the sun looks larger than any other star.

The sun is many times larger than Earth.

The sun gives heat and light to everything on Earth. Plants, animals, and people need the sun.

Earth is one of the eight planets. The planets travel, or revolve, around the sun. The path each planet takes is an orbit.

Observe how each planet is a different distance from the sun.

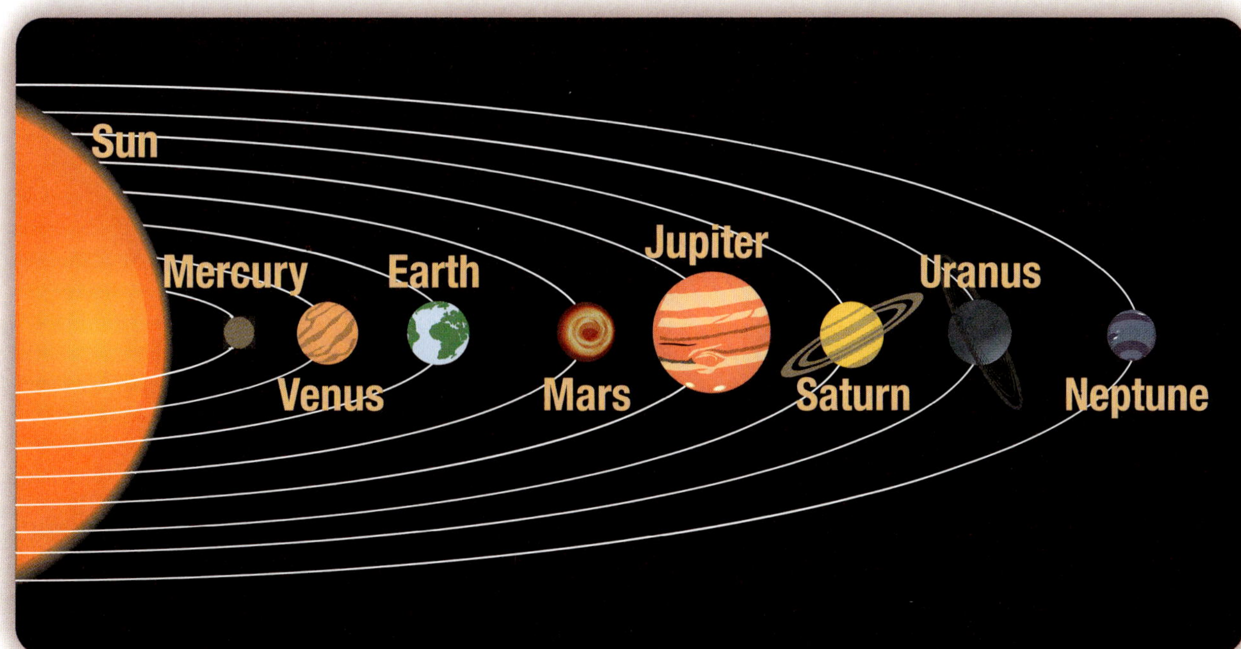

Which planet is closest to the sun? farthest from? Which planet is the largest? the smallest?

Which planet would you like to visit? Why?

God made two great lights—the greater light to govern the day and the lesser light to govern the night. He also made the stars. Genesis 1:16

Sun, Planets, and Moon

What do you see when you look at the moon from Earth?

Earth has one moon. It orbits Earth.

The moon cannot make its own light. The part of the moon you see has light from the sun shining on it.

The moon is round. When you see a circle, you see a full moon. You may see different shapes called phases. You may not be able to see the moon at all.

There is no air or water on the moon. There are no living things.

The moon has deep holes, mountains, and flat places. The flat places are the dark spots that you see.

Apollo 11 was a mission to the moon. The first person to walk on the moon was astronaut Neil Armstrong.

Lesson Review

Fill in the circle beside each answer.

1. What is the closest star to Earth?

   ○ planet ○ sun ○ moon

2. Are there living things on the moon?

   ○ yes ○ no

Have you ever tried to count the stars?

At night, you can see many stars in the sky. Stars are hot balls of gas that shine.

Stars look small because they are very far away. The sun is a star. It looks larger because it is not as far away.

In Bible times, people imagined there were lines connecting stars. They pretended to see pictures in the sky.

A group of stars that forms a pattern is called a **constellation**.

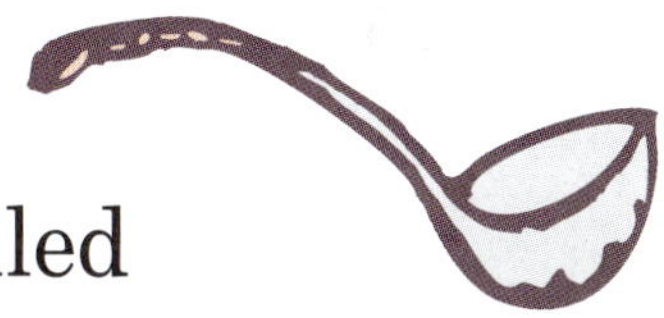

Can you see why this group of stars is called the Big Dipper?

Big Dipper

Using imaginary lines, can you find the star picture of a bear? Draw lines to connect the stars. Do you see the Big Dipper?

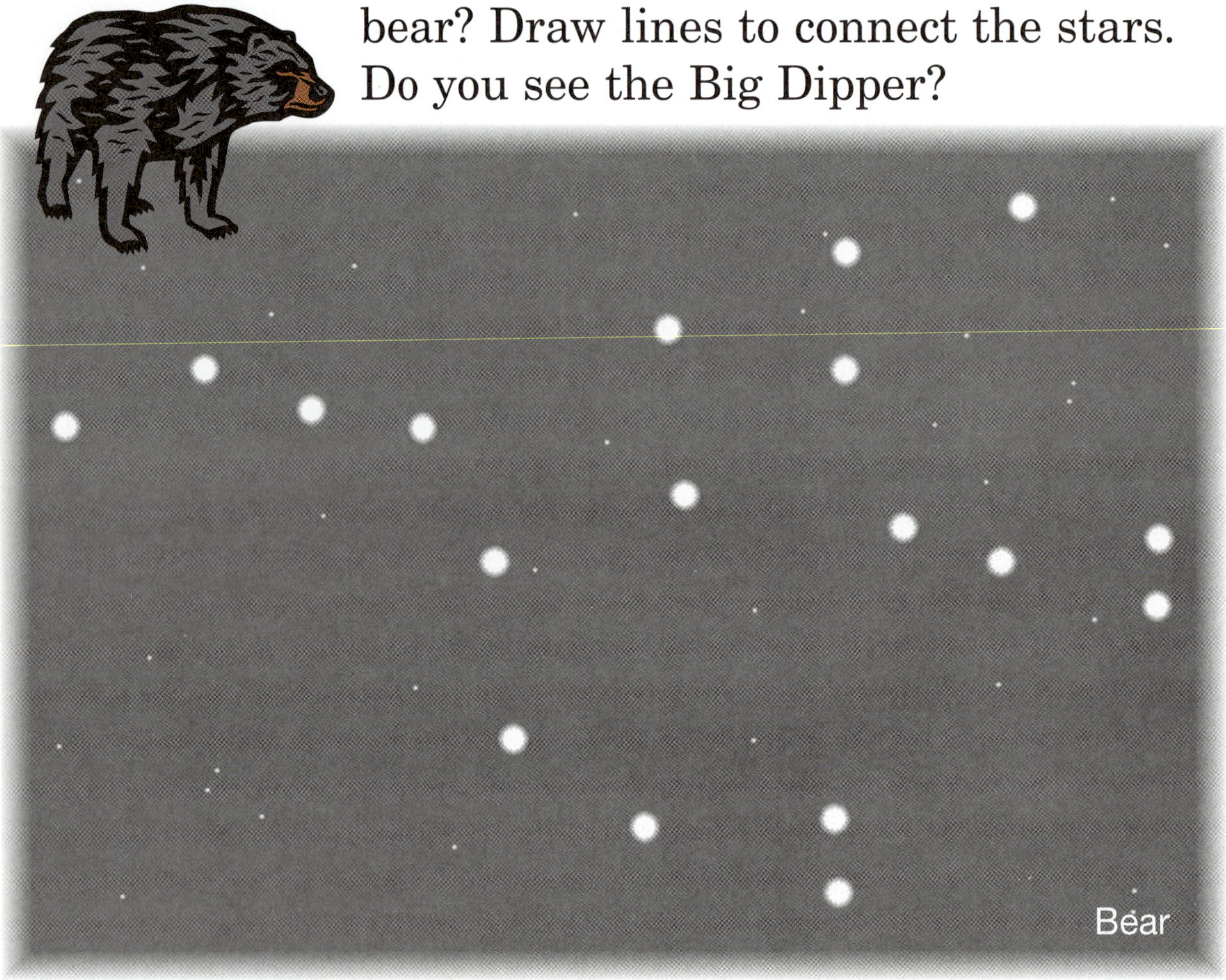

He is the Maker of the Bear and Orion . . . and the constellations of the south. Job 9:9

Try making a star picture.

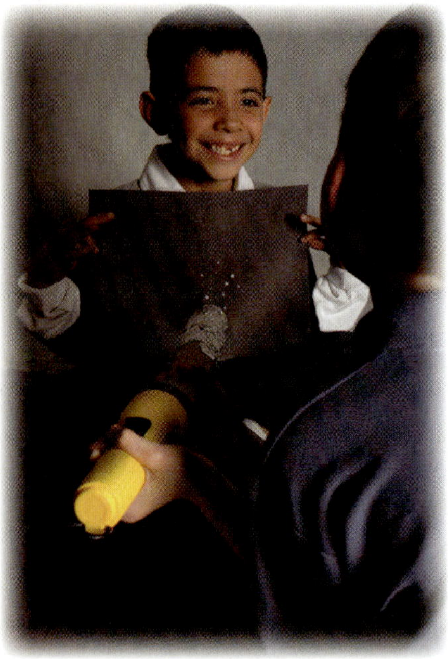

- Tape aluminum foil over one end of a paper towel tube.
- Think of a pattern for a picture.
- Use a pencil to make holes.
- Shine a flashlight through the other end of the tube.
- Ask someone to tell what the star picture is.

In 1926, Dr. Goddard's rocket was sent up. Other rockets and missiles continued to help scientists and those at war.

A time line shows the order that events took place. What does this time line show?

1950 — 1960 — 1970 — 1980 — 1990 — 2000

1957
Sputnik I, a satellite, makes the first orbit of Earth.

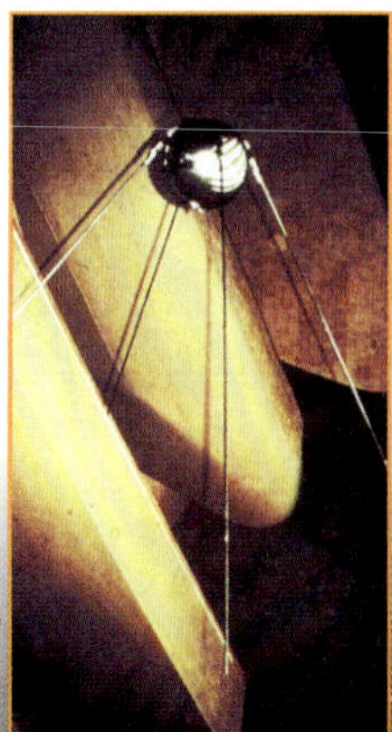

1969
Neil Armstrong is the first person to walk on the moon.

1971
The first spacecraft to orbit another planet, Mars, is *Mariner 9*.

1981
The space shuttle missions begin.

1998
The *International Space Station* project begins.

Pretend you are in this space shuttle. Tell where you are going.

Draw what you see from the space shuttle window.

Tech Connect

How does exploring space help people?

People have many things because scientists study and experiment in space.

People have cell phones to talk with one another.

The Fisher space pen will write upside down, in hot and cold temperatures, and underwater.

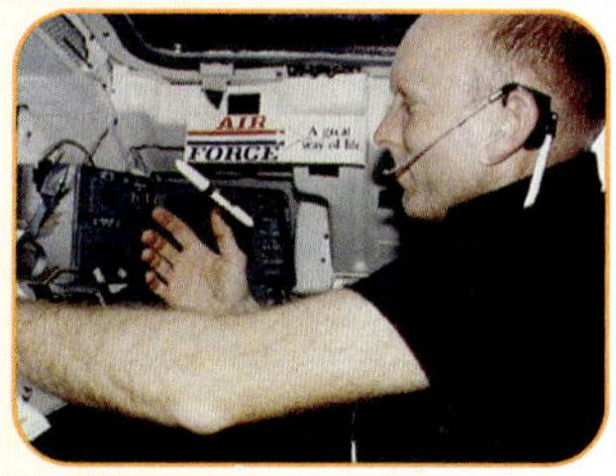

Dehydrated food may be eaten by fishermen and hikers.

Doppler radar is used to predict the weather.

A skier can enjoy boots that bend and fogless goggles.

Underline the word in each sentence above that tells what people have because of exploring space.

Tech-Connect

Experiment

Materials
3 inch by 5 inch card

Skills		
predict	experiment	observe

Problem Can you place the space shuttle in space without cutting it out?

Predict Make a guess about the problem. Circle *yes* or *no*.

Plan How can you test this prediction?

Experiment Place the card on its long side between the space shuttle and space. Touch the card with the tip of your nose. Keep both eyes open.

Observe Can you observe the space shuttle in space? Circle *yes* or *no*.

Conclusion The space shuttle is in ______________________.

Lord, you have made the heavens and the earth by your great power. . . . Nothing is too hard for you. Jeremiah 32:17b

Draw a line to the word that completes each sentence.

1. Planets revolve around the _____.
2. Different shapes of the moon are _____.
3. A _____ makes things that are far away appear closer.

- telescope 
- sun 
- phases 

Write *yes* or *no*.

4. Can you see the atmosphere? ____________________

5. Does the atmosphere protect Earth? ____________________

Circle the word that answers each question.

6. Are the people who live here having *day* or *night*?

7. Is it *morning* or *noon* in this picture?

8. Circle each object that belongs in space.

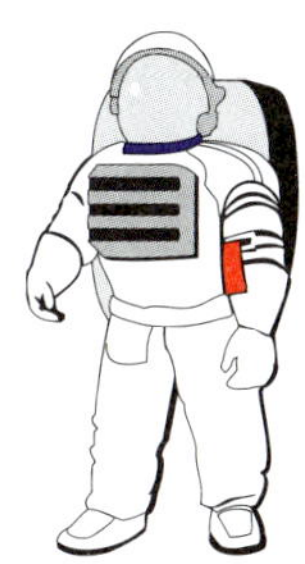

 Purposeful Design Publications

Chapter 13 Review

Read the creation web.

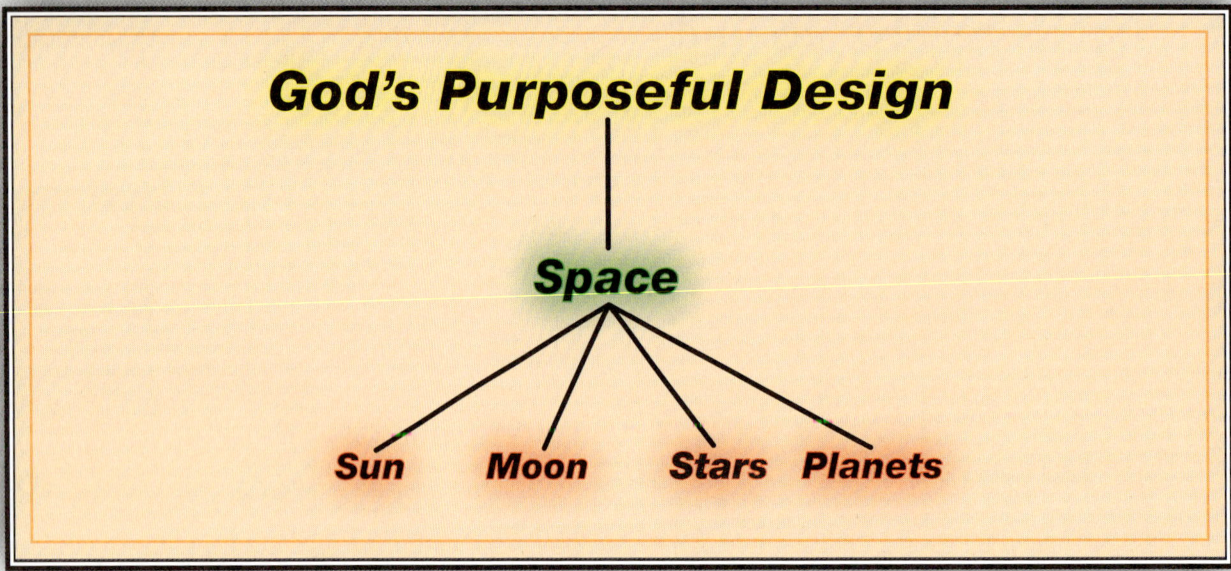

Journal about Space

Would you like to be an astronaut? Explain.